Encyclopedia of Gas Chromatography: New Developments in Diverse Fields

Volume I

Encyclopedia of Gas Chromatography: New Developments in Diverse Fields

Volume I

Edited by **Carol Evans**

New York

Published by NY Research Press,
23 West, 55th Street, Suite 816,
New York, NY 10019, USA
www.nyresearchpress.com

Encyclopedia of Gas Chromatography: New Developments in Diverse Fields
Volume I
Edited by Carol Evans

International Standard Book Number: 978-1-63238-128-6 (Hardback)

Printed in the United States of America.

Contents

Preface

This book is a collection of studies on current advancements and growth in the field of gas chromatography and its applications. Diverse uses varying from basic biological applications to industrial uses have been overviewed in this book. Analysis of current advancements in chromatographic columns, microextraction techniques, derivatisation techniques and pyrolysis techniques has been presented in detail. It also comprises of numerous attributes of fundamental chromatography processes and is a valued resource for both budding and expert chromatographers. Latest advancements like inverse and multidimensional chromatography have also been dealt with. Studies of various toxicants, petroleum hydrocarbons and naturally occurring components have also been presented in the book. The various topics have been addressed by masters from diverse spheres and supplemented by easy-to-understand figures and tables. It has been specially designed for chemists and other practitioners working in the domains related to gas chromatography.

This book is a comprehensive compilation of works of different researchers from varied parts of the world. It includes valuable experiences of the researchers with the sole objective of providing the readers (learners) with a proper knowledge of the concerned field. This book will be beneficial in evoking inspiration and enhancing the knowledge of the interested readers.

In the end, I would like to extend my heartiest thanks to the authors who worked with great determination on their chapters. I also appreciate the publisher's support in the course of the book. I would also like to deeply acknowledge my family who stood by me as a source of inspiration during the project.

Editor

Part 1

New Development in Basic Chromatographic and Extraction Techniques

1

Hyphenated Techniques in Gas Chromatography

Xinghua Guo* and Ernst Lankmayr
Institute of Analytical Chemistry and Food Chemistry, Graz University of Technology, Austria

1. Introduction

Hyphenated gas chromatography (GC) (Chaturvedi & Nanda, 2010; Coleman III & Gordon, 2006) mainly refers to the coupling of the high-performance separation technique of gas chromatography with 1) information-rich and sophisticated GC detectors which otherwise can be mostly operated as a stand-alone instrument for chemical analysis, and 2) automated online sample preparation systems. The term "hyphenation" was first adapted by Hirschfeld in 1980 to describe a possible combination of two or more instrumental analytical methods in a single run (Hirschfeld, 1980). It is of course not the case that you can couple GC to any detection systems, although many GC hyphenations have been investigated and / or implemented as to be discussed in this chapter. The aim of this coupling is obviously to obtain an information-rich detection for both identification and quantification compared to that with a simple detector (such as thermal-conductivity detection (TCD), flame-ionization detection (FID) and electron-capture detector (ECD), etc.) for a GC system.

According to the detection mechanism, information-rich detectors can be mainly classified as 1) detection based on molecular mass spectrometry, 2) detection based on molecular spectroscopy such as Fourier-Transform infrared (FTIR) and nuclear magnetic resonance (NMR) spectroscopy, and 3) detection based on atomic spectroscopy (elemental analysis) by coupling with such as inductively-coupled plasma (ICP)-MS, atomic absorption spectroscopy (AAS) and atomic emission spectroscopy (AES), respectively. In addition to these hyphenations mentioned above and which are mounted after a gas chromatograph, it can also include automated online sample preparation systems before a GC system such as static headspace (HS), dynamic headspace, large volume injection (LVI) and solid-phase microextraction (SPME). One of the recent developments is the hyphenation of GC with human beings – so called GC-Olfactometry (GC-O) or GC-Sniffer (Friedrich & Acree, 1998). Hyphenated gas chromatography also include coupling of gas chromatographs orthogonally - multidimensional gas chromatography (MDGC or GC×GC).

2. Hyphenated techniques in gas chromatography

This chapter provides a general overview of the current hyphenated GC techniques with focus on commonly applied GC-MS, GC-FTIR (GC-NMR) for detection of molecular

---

* Corresponding Author

analytes as well as GC-AAS and GC-AES coupling for elemental analysis. Emphasis will be given to cover various GC-MS techniques including ionization methods, MS analysers, tandem MS detection and data interpretation. For more comprehensive overview on their applications, readers are directed to other chapters in this book or other dedicated volumes (Grob & Barry, 2004; Message, 1984; Jaeger, 1987; Niessen, 2001).

2.1 Gas Chromatography-Mass Spectrometry (GC-MS)

In 1957, Holmes and Morrell (Holmes & Morrell, 1957) demonstrated the first coupling of gas chromatography with mass spectrometry shortly after the development of gas-liquid chromatography (James & Martin, 1952) and organic mass spectrometry (Gohlke & McLafferty, 1993). Years later, improved GC-MS instruments were commercialized with the development of computer-controlled quadrupole mass spectrometer for fast acquisition to accommodate the separation in gas chromatograph. Since then, its applications in various areas of sciences has made it a routine method of choice for (bio)organic analysis (Kuhara, 2005).

As its name suggested, a GC-MS instrument is composed of at least the following two major building blocks: a gas chromatograph and a mass spectrometer. GC-MS separates chemical mixtures into individual components (using a gas chromatograph) and identifies / quantifies the components at a molecular level (using a MS detector). It is one of the most accurate and efficient tools for analyzing volatile organic samples.

The separation occurs in the gas chromatographic column (such as capillary) when vaporized analytes are carried through by the inert heated mobile phase (so-called carrier gas such as helium). The driven force for the separation is the distinguishable interactions of analytes with the stationary phase (liquid thin layer coating on the inner wall of the column or solid sorbent packed in the column) and the mobile phase respectively. For gas-liquid chromatography, it depends on the column's dimensions (length, diameter, film thickness), type of carrier gas, column temperature (gradient) as well as the properties of the stationary phase (e.g. alkylpolysiloxane). The differences in the boiling points and other chemical properties between different molecules in a mixture will separate the components while the sample travels through the length of the column. The analytes spend different time (called retention time) to come out of (elute from) the GC column due to their different adsorption on the stationary phase (of a packed column in gas-solid chromatography) or different partition between the mobile phase (carrier gas) and the stationary phase (of a capillary column in gas-liquid chromatography) respectively.

As the separated substances emerge from the column opening, they flow further into the MS through an interface. This is followed by ionization, mass-analysis and detection of mass-to-charge ratios of ions generated from each analyte by the mass spectrometer. Dependent on the ionization modes, the ionization interface for GC-MS can not only ionize the analytes but also break them into ionized fragments, and also detect these fragments together with the molecular ions such as, in positive mode, radical cations using electron impact ionization (EI) or (de-)protonated molecules using chemical ionization (CI). All ions from an analyte together (molecular ions or fragment ions) form a fingerprint mass spectrum, which is unique for this analyte. A "library" of known mass spectra acquired under a standard condition (for instance, 70-eV EI), covering several hundreds of thousand compounds, is stored on a computer. Mostly a database search can identify an unknown component rapidly in GC-MS. Mass spectrometry is a powerful tool in instrumental analytical chemistry because it provides more information about the composition and structure of a substance from less sample than any other analytical technique, and is considered as the only definitive analytical detector among all available GC detection techniques.

The combination of the two essential components, gas chromatograph and mass spectrometer in a GC-MS, allows a much accurate chemical identification than either technique used separately. It is not easy to make an accurate identification of a particular molecule by gas chromatography or mass spectrometry alone, since they may require a very pure sample or standard. Sometimes two different analytes can also have a similar pattern of ionized fragments in their standard mass spectra. Combining the two processes reduces the possibility of identification error. It is possible that the two different analytes will be separated from each other by their characteristic retention times in GC and two co-eluting compounds then have different molecular or fragment masses in MS. In both cases, a combined GC-MS can detect them separately.

2.1.1 Ionization techniques and interfaces

The carrier gas that comes out of a GC column is primarily a pressurized gas with a flow about mL/min for capillary columns and up to 150 mL/min for packed columns. In contrast, ionization, ion transmission, separation and detection in mass spectrometer are all carried out under high vacuum system at approximately 10^{-4} Pa (10^{-6} Torr). For this reason, sufficient pumping power is always required at the interface region of a GC-MS in order to provide a compatible condition for the coupling. When the analytes travel through the length of the column, pass through the transfer line and enter the mass spectrometer they can be ionized by several methods. Fragmentation can occur alongside ionization too. After ion separation by mass analyser, they will be detected, usually by an electron multiplier diode, which essentially turns the ionized analytes/fragments into an electrical signal.

The two well-accepted standard types of the ionization techniques in GC-MS are the prevalent electron impact ionization (EI) and the alternative chemical ionization (CI) in either positive or negative modes.

Electron impact (EI) ionization:

The EI source is an approximately one cubic centimetre device, which is located in the ion source housing as shown in Figure 1. The ion source is open to allow for the maximum conductance of gas from the ion source into the source housing and then into the high-vacuum pumping system. The EI source is fitted with a pair of permanent magnets that cause the electron beam to move in a three-dimensional helical path, which increases the probability of the interaction between an electron and an analyte molecule. Even under this condition, only about 0.01~0.001% of the analyte molecules are actually ionized (Watson, 1997). During the EI ionization, the vaporized molecules enter into the MS ion source where they are bombarded with free electrons emitted from a heated filament (such as rhenium). The kinetically activated electrons (70-eV) collide with the molecules, causing the molecule to be ionized and fragmented in a characteristic and reproducible way.

Due to the very light weight of the electrons (1/1837 that of the mass of a proton or neutron), one collision between an electron and a molecule would bring insufficient internal energy to the molecule to become ionized. The molecule's internal energy is increased by the interaction of the electron cloud after a collision cascade. The energized molecule, wanting to descend to a lower energy state, will then expel one of its electrons. The result is an odd-electron species called a radical cation, which is the molecular ion and has the same integer mass as the analyte molecule. Some of the molecular ions will remain intact and pass through the *m/z* analyser and be detected. The molecular weight of the analyte is represented by the molecular ion peak in the mass spectrum if there are any. The single-charge molecular ion has the same nominal mass of the molecule, but the accurate mass is differed by that of an electron. Most molecular ions will undergo unimolecular decomposition to produce fragment ions.

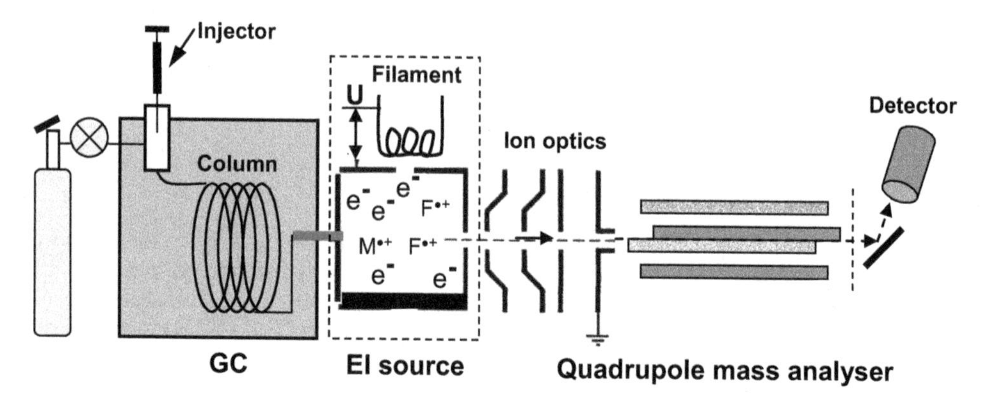

Fig. 1. Schematic drawing of a GC-MS with EI ionization and quadrupole mass analyser.

The reason to apply the 70 eV as the standard ionization energy of electrons is to build a spectrum library with the standard mass spectra and subsequently to perform the MS database searching for unknown identification. The molecular fragmentation pattern is dependant upon the electron energy applied to the system. The excessive energy of electrons will induce fragmentation of molecular ions and results in an informative fingerprint mass spectrum of an analyte (Watson, 1997). The use of the standard ionization energy facilitates comparison of generated spectra with library spectra using manufacturer-supplied software or software developed by the National Institute of Standards (NIST-USA). The two main commercial mass spectra libraries for general purpose are the NIST Library and the Wiley Library. In addition, several small libraries containing a specific class of compounds have also been developed by individual manufacturers or research institutions. This includes libraries for pesticides, drugs, flavour and fragrance, metabolites, forensic toxicological compounds and volatile organic compounds (VOC), just to name a few. Spectral library searches employ matching algorithms such as probability based matching (McLafferty et al., 1974) and dot-product (Stein & Scott, 1994) matching that are used with methods of analysis written by many method standardization agencies. EI is such an energetic process, in some cases, there is often no molecular ion peak left in the resulting mass spectrum. Since the molecular ion can be a very useful structure information for identification, this is sometimes a drawback compared to other soft ionization such as chemical ionization, field ionization (FI), or field desorption ionization (FDI), where the desired (pseudo) molecular ion will be generated and detected. After ionization, the stable ions (which remain without dissociation during its flight from the ion source to the detector) will be pushed out of the EI source by an electrode plate with the same charge polarity, called the repeller. As the ions exit the source, they are accelerated / transmitted into the mass analyser of the mass spectrometer.

Chemical ionization (CI):

CI is a less energetic process than EI. In the latter case, the ionizing electrons accelerated to have some kinetic energy collide directly with gas-phase analyte molecules to result in ionization accompanied by simultaneous fragmentation. However, CI is a low energetic ionization technique generating pseudo molecular ions such as $[M+H]^+$ rather than the conventional $M^{\bullet +}$, and inducing less fragmentation. This low energy usually leads to a simpler mass spectrum showing easily identifiable molecular weight information. In CI a reagent gas such as methane, ammonia or isobutane is introduced into the relatively closed ion source of a mass spectrometer at a pressure of about 1 Torr. Inside the ion source, the reagent gases (and the subsequent reagent ions) are present in large excess compared to the analyte. Depending on the technique (positive CI or negative CI) chosen, the pre-existing reagent gas in the ion source will interact with the electrons preferentially first to ionize the reagent gas to produce some reagent ions such as CH_5^+ and NH_4^+ in the positive mode and OH^- in the negative mode through self-CI of the reagent gas. The resultant collisions and ion/molecule reactions with other reagent gas molecules will create an ionization plasma. When analytes are eluted (from GC) or evaporated (direct inlet), positive and negative ions of the analyte are formed by chemical reactions such as proton-transfer, proton-subtraction, adduct formation and charge transfer reactions (de Hoffmann & Stroobant, 2003), where the proton transfer to produce $[M+H]^+$ ions is predominate during the ionization. The energetics of the proton transfer is controlled by using different reagent gases. Methane is the strongest proton donor commonly used with a proton affinity (PA) of 5.7 eV, followed by isobutane (PA 8.5 eV) and ammonia (PA 9.0 eV) for 'softer' ionization. As mentioned above, in this soft ionization, the main benefit is that the (pseudo) molecular ions closely corresponding to the molecular weight of the analyte of interest are produced. This not only makes the follow-up identification of the molecular ions easier but also allows ionization of some thermal labile analytes.

The sensitivity and selectivity of a mass spectrometer in a hyphenated GC largely depend on the interfacing technique: the ion source and the ionization mode. The sensitivity is related to the ionization efficiency and the selectivity is primarily to the ionization mode. Both sensitivity and selectivity can differ for different classes of compounds. A proper choice of an ionization technique is a key step for a successful GC-MS method development. In most case of EI ionization, the positive mode is much preferred due to its capability efficiently to ionize most analytes and to induce sufficient fragmentation to build up a database with positive mass spectra. The sensitivity in the negative EI mode will be much lower. However, the fragmentation processes in EI can be so extensive that the molecular ion is absent in the mass spectrum of some compounds. As a result, the useful molecular weight information is lost, which is indeed a disadvantage of this ionization technique (Karasek & Clement, 1988; Ong & Hites, 1994). In this aspect, chemical ionization is a reasonable alternative and supplemental to EI as a soft ionization method. Due to the moderate energy transmission to the analytes, the pseudo molecular ions are formed in CI and these even-electron ions are more stable than the odd-electron ions produced with EI

(Harrison, 1992). Unlike EI, both positive chemical ionization (PCI) and negative chemical ionization (NCI) are equally used according to their specific features. The positive mode is best suitable for hydroxyl group-containing alcohols and sugars as well as basic amino compounds. The sensitivity of PCI-MS is comparable with low-resolution EI-MS for most compounds. On the other hand, the negative mode is widely used in environmental analysis because it is highly selective and sensitive to, for instance, organochlorine (halogen-containing) and acidic group-containing compounds (Karasek & Clement, 1988). For this, the sensitivity of the NCI is significantly better compared with that of the PCI. As a result, the NCI has also been equally developed in the sense of the choices of the reagent gases and their combination (Chernetsova et al., 2002). Furthermore, the application of an alternating or simultaneous EI and CI in GC-EI/CI-MS has also been reported (Arsenault et al., 1971 & Hunt et al., 1976). Although it has not been popularly implemented in the commercial GC-MS systems, the advantages of obtaining informative mass spectra in trace analysis of samples with limited quantities are very promising. In addition to the above-mentioned EI and CI ionization interfaces, where the ionization occurs in the vacuum of a mass spectrometer, a recent development has also demonstrated an interfacing technique at the atmosphere. That is the coupling of a GC with a mass spectrometer with the so-called Direct Analysis in Real Time (DART) (Cody et al., 2005 & Cody 2008), where the absence of vacuum interface, electron filament and CI reagent gases offers a robust interfacing technique for rapid analysis.

2.1.2 Molecular weight, molecular ion, exact mass and isotope distribution

For a singly charged analyte, the mass-to-charge ratio of its molecular ion indicates the molecular weight (*a.m.u.*) of the analyte. The molecular ion can be a radical cation $[M]^{\bullet +}$ (positive mode) or a radical anion $[M]^{\bullet -}$ (negative mode), derived from the neutral molecule by kicking off / attaching one electron. According to the IUPAC definition (McNaught & Wilkinson, 1997), the molecular mass (*Mr*) is the ratio of the mass of one molecule of that analyte to the unified atomic mass unit *u* (equal to 1/12 the mass of one atom of ^{12}C). It is also called molecular weight (*MW*) or relative molar mass (*Mr*). Knowing the fact that each isotope ion (rather than their average) of an analyte is detected separately in MS mass analyser, one should pay attention to the different definitions of molecular masses in mass spectrometry and GC-MS.

The nominal mass of an ion is calculated using the integer mass by ignoring the mass defect of the most abundant isotope of each element (Yergey et al., 1983). This is equivalent as summing the numbers of protons and neutrons in all constituent atoms. For example, for atoms H = 1, C = 12, O = 16, etc. Nominal mass is also called mass number. To assign elemental composition of an ion in low resolution MS, the nominal mass is often used, which enable to count the numbers of each constituent elements. However, if the sum of the masses of the atoms in a molecule using the most abundant isotope (instead of the isotopic average mass) for each element is calculated (Goraczko, 2005), the monoisotopic mass is obtained (McNaught & Wilkinson, 1997). Monoisotopic mass is also expressed in unified atomic mass units (a.m.u.) or Daltons (Da). For typical small molecule organic compounds with elements C, H, O, N etc., this is the mass that one can measure with a mass spectrometer and it often refers to the lightest isotope in the isotopic distribution. However, for a molecule containing special elements such as B, Fe and Ar, etc., the most abundant isotope is not the lightest one any more. Another important and useful term is the exact mass, which is obtained by summing the exact masses of the individual isotopes of the molecule (Sparkman, 2006). According to this definition, several exact masses can be calculated for one chemical formula depending on the constituent isotopes. However, in practical mass spectrometry, exact mass refers to that corresponding to the monoisotopic mass, which is the sum of the exact masses of the most abundant isotopes for all atoms of the molecule. This is the peak one can measure with high resolution mass spectrometry. This results in measured accurate mass (Sparkman, 2006), which is an experimentally determined mass that allows the elemental composition to be determined either using the accurate mass mode of a high resolution mass spectrometer (Grange et al., 2005). For molecules with mass below 200 u, a 5 ppm accuracy is sufficient to uniquely determine the elemental composition. And at *m/z* 750, an error of 0.018 ppm would be required to eliminate all extraneous possibilities (Gross, 1994). When the exact mass for all constituent isotopes (atoms or molecules) are known, the average mass of a molecule can be calculated. That is the sum of the average atomic masses of the constituent elements. However, one can never measure the average mass of a compound using mass spectrometry directly, but that of its individual isotopes.

Some molecular ions can be obtained using EI ionization. If not, the use of soft-ionization techniques (CI, FI) would help to facilitate identification of the molecular ion. Accurate masses can be determined using high resolution instruments such as magnet/sectors, reflectron TOFs (see Section 2.1.4), Fourier-Transform ion cyclotron resonance (FT-ICR) (Marshall et al., 1998) and Orbitrap (Makarov, 2000) mass spectrometers. Because of isotopes of some elements, molecular ions in mass spectrometry are often shown as a distribution of isotopes. The isotopic distribution can be calculated easily with some programs freely available and allows you to predict or confirm the masses and abundances of the isotopes for a given chemical formula. Isotopic patterns observed are helpful for predicting the appropriate number of special elements (e.g., Cl, Br and S) or even C numbers. When a Library mass spectrum is not available, this will be one of the important means for unknown identification. The elucidation process normally requires not only the combined use of hard and soft ionization techniques but also tandem MS experiments to obtain information of fragments for data interpretation. In some cases, an elemental composition might be proposed for the unknown based on isotope patterns and accurate mass measurements of the molecular and fragment ions (Hernández et al, 2011) (see Section 2.1.4).

2.1.3 Typical mass analyzers and MS detectors in GC-MS

MS instruments with different mass analysers, e.g. magnet/electric sectors, quadrupoles (linear), ion traps (Paul traps & quadrupole linear ion trap), FT-ICR and time-of-flight (TOF) mass analysers have all been implemented for the coupling of GC and MS. Since its conception, linear quadrupoles have been dominating the GC-MS applications.

Quadrupole mass analyser:

A single-stage linear quadrupole mass analyser can be considered as a mass filter and it consists of four hyperbolic metal rods placed parallel in a radial array. A pair of the opposite rods have a potential of $(U+V\cos(\omega t))$ and the other pair have a potential of $-(U+V\cos(\omega t))$. An appropriate combination of direct current (DC) and radio frequency (RF, ~ 1 MHz) electric field applied to the four rods induces an oscillatory motion of ions guided axially into the assembly by means of a low accelerating potential. The oscillating trajectories are mass dependent and ions with one particular mass-to-charge (*m/z*) ratio can be transmitted toward the detector when a stable trajectory through the rods is obtained. At static DC and RF values, this device is a mass filter to allow the ions at this *m/z* pass through and all others will be deflected. Ions of different *m/z* can be consecutively transmitted by the quadrupole mass filter toward the detector when the DC and RF potentials are swept at a constant ratio and oscillation frequencies (de Hoffmann & Stroobant, 2003; Santos & Galceran, 2003).

A single quadrupole mass analyser can be operated in either the full-spectrum scan mode or the selected ion monitoring (SIM) mode. In the scan mode, the ions of a certain *m/z* range will pass through the quadrupole sequentially while scanning the DC and RF potentials to the detector. The advantage is that this allows to record a full mass spectrum containing the information from molecular ion to fragments, which further enables a library searching to compare with the standard mass spectra for unknown identification. It is common to require a scanning of the analyser from about 50 a.m.u. to exclude the background ions from residue air, carrier gases, and CI reagent gases. However, especially in the earlier years, the slow scan speed (250-500 a.m.u. per second) for a quadrupole analyser was an important limiting factor and it is in contrary with the high resolution (narrow peaks about a few seconds wide) of a capillary GC separation. Apparently, a longer dwell time is essentially required to obtain a more sensitive detection. Therefore, the SIM mode is preferred for a selective and sensitive detection in quantification of known analytes, where the DC and RF are set to pre-defined values according to the *m/z* of the analyte (molecular ion or specific fragments) for up to hundred milliseconds. Linear quadrupoles are the most widely used mass analysers in GC–MS, mainly because they make it possible to obtain high sensitivity, good qualitative information and adequate quantitative results with relatively low maintenance. Generally, these instruments are characterised by a bench-top configuration with unit mass resolution and both electron impact and chemical ionisation techniques. The relatively low cost, compactness, moderate vacuum requirements and the simplicity of operation make quadrupole mass spectrometers the most popular mass analyser for GC-MS. The continuous and significant improvements in scanning speed (to allow tandem MS detection), sensitivity and detection limits has also been observed in the past decade. New developments have been implemented in GC–MS instrumentations based on quadrupole technology with regard to the stability of mass calibration, the fast scan-speed with a higher sensitivity. It is now also possible to work simultaneously with full-scan and selected ion monitoring (SIM) modes in a single GC-MS run.

Ion trap (IT) mass analyser:

An ion trap mass spectrometer uses a combination of electric or magnetic fields to capture and store ions in a vacuum chamber. According to its principles of operation, it can be specified as quadrupole ion traps (Paul trap) (Paul & Steinwedel, 1953), quadrupole linear ion traps (Schwartz et al., 2002), FTICR-MS (Penning trap) (Marshall et al., 1998) and Orbitraps (Kington traps) (Kington, 1923; Makarov, 2000), respectively. However, the coupling to GC has been dominated by quadrupole mass analysers and quadrupole ion traps. Due to the applicability and other limitations, only very recently, there have been reports on implementing FTICR-MS (Szulejko & Solouki, 2002) and Orbitrap-MS (Peterson et al., 2010) as detectors in GC-MS.

The three-dimensional Paul ion trap (IT) is composed of a central hyperbolic ring electrode located between two symmetrical hyperbolic end-cap electrodes. The geometry of the device is described by the relationship $r_0^2 = 2\times z_0^2$, where r_0 is defined as the radius of the ring electrode and $2z_0$ is the distance between the end-cap electrodes. The value r_0 is used to specify a trap, which ranges about 1-25 mm. During the ion trapping, an auxiliary oscillating potential of low amplitude is applied across the end-cap electrodes while a RF potential of ~ 1 MHz is applied to the ring electrode. As the amplitude of the RF potential is increased, the ions become more kinetically energetic and they develop unstable trajectories (excited for dissociation or ejected to detector). One of the significant advantages of ion traps compared to the above-mentioned linear quadrupole mass analysers is their high sensitivity in full scan mode. Based on this, many qualitative and quantitative applications have been reported (Allchin et al., 1999 & Sarrion et al., 2000). Beside the conventional full scan mode of operation, an important advancement by application of ion trap in GC-MS has been the capability of an ion trap mass analyser to perform tandem mass spectrometric investigations of ions of interest in collision-induced dissociation (CID) for ion structure elucidation (Plomley et al., 2000) (see the text below for tandem mass spectrometers).

For both linear quadrupole and 3-D mass spectrometers, after mass-resolved ions passing the mass analyser, they continue to travel to the ion detector. It is essential in GC-MS to have fast responding detector with a broad range of magnification. The most popular detector employed is the electron multiplier composed of a series of dynodes. The ions collide with the first one to generate primary electrons. Due to the increasing high voltage (kV) between the dynodes, the electrons further collide with the next dynode and generate more electrons, and so on. Therefore, a small ion current can be finally amplified into a huge signal. Another commonly used detector is the photomultiplier. The ions collide with a phosphor-coated target and are converted into photons that are subsequently magnified and detected. Typically, these detectors are operated at lower voltages (400-700 V) and, therefore, have a longer lifetime than the electron multipliers (using high kV voltages) (Grob, 1995).

Time-of-flight (TOF) mass analyser:

Time-of-flight mass spectrometry (TOF-MS) is based on a simple mass separation principle in which the m/z of an ion is determined by a measurement of its flight time over a known distance (Cotter, 1994; Stephens, 1946). Pulsed ions are initially accelerated by means of a constant homogeneous electrostatic field of known strength to have the same kinetic energy (given they have the same charge). Therefore, the square of the velocity of an ion is reversely proportional to its m/z, $E_k = (1/2)mv^2$ and the time of arrival t at a detector directly indicates its mass (Equation 1).

$$t = \mathrm{d}\times(m/\mathrm{q})^{1/2}\times(2\mathrm{U})^{-1/2} \quad (1)$$

where d is the length of the flight tube, m is the mass of the ion, q is the charge, U is the electric potential difference used for the acceleration. The arrival time t can be measured using a transient digitizer or time to digital converter, and is about milliseconds. A TOF mass analyser has theoretically no mass limit for detection and a high sensitivity. However, the spread of kinetic energies of the accelerated ions can lead to different arrival time for the ions with the same m/z and result in low mass resolution. This can be overcomed by applying a reflectron flight path (Mamyrin et al., 1973) rather than a linear one to refocus the ions with the same m/z value to arrive at the detector simultaneously.

In TOF-MS, ions must be sampled in pulses. This fits well with a laser ionization source. However, when EI or CI are used for GC-MS coupling, a continuous ion beam is produced. Therefore, in most GC-TOF-MS instrumentations, a pulse of an appropriate voltage is applied to deflect and accelerate a bunch of the ions in the orthogonal direction to their initial flight path (so-called oa-TOF-MS). One of the important advantages of TOF-MS as a GC detector is its capability of producing mass spectra within a very short time (a few milliseconds), with high sensitivity. Furthermore, its high mass accuracy (errors in low ppm) has made it an alternative to accurate mass GC-MS using magnet / sector instruments (see Section 2.1.4). In both GC-MS and LC-MS, hybridized TOF-MS instruments with quadrupoles (Q-TOF-MS) (Chernushevich, 2001), ion trap (IT-TOF-MS), and even another TOF (such as TOF/TOF-MS) have all been developed and found more applications than TOF-MS alone (Vestal & Campbell, 2005).

Tandem mass spectrometer as a mass analyser:

Tandem mass spectrometer refers to an arrangement of mass analysers in which ions are subjected to two or more sequential stages of analysis (which may be separated spatially or temporally). The study of ions involving two stages of mass analysis has been termed mass spectrometry/mass spectrometry (MS/MS) (Todd, 1991). For GC applications, it includes mainly triple quadrupole, 3D-ion trap and linear quadrupole ion trap tandem mass spectrometers and only very recently also FTICR (Szulejko & Solouki, 2002 & Solouki et al., 2004) and Orbitrap-MS (Peterson et al., 2010) as detectors in GC-MS.

Triple quadrupole MS/MS involves two quadrupole analysers mounted in a series but operating simultaneously (either as a mass filter or scanner) and with a collision cell between two mass analysers. Since the analysers can be scanned or set to static individually, types of operations include product ion scan, precursor ion scan and neutral loss scan. Product ion scan enables ion structure studies. This is realised by isolation of ions of interest according to their *m/z* using first quadrupole analyser. Then dissociation of the ions after kinetic-energetically activated occurs in the collision cell filled with an inert collision gas. Finally, the product ions will be mass-analysed using the third quadrupole. The fragment information provides further insights into ion structures or functional groups. This can be used to confirm structures of components in question when it is compared with a reference standard or to deduce ion structures for new chemical entities in qualitative identification. Some mass spectra databases include even a library of product ion spectra for some selected compounds to assistant the identification. Furthermore, the precursor ion scan can be utilised to study multiple precursor ions of a certain fragment ion. This is very useful to study a class of compounds producing a common charged fragment during collision-induced dissociation (CID) in the collision cell. For instance, the protonated 1,2-benzenedicarboxylic acid anhydride at *m/z* 149 derived from almost all corresponding phthalate esters can be used as such an ion and, during the precursor scan, all related phthalates will be found. On the other hand, neutral loss scan can be used to study a class of compounds showing a common neutral molecule loss during CID. For instance, the loss of a CO_2 from most deprotonated carboxylic acids or H_2O from protonated alcohols can be applied for this purpose. For dissociation reactions found in either product scan or precursor ion scans, a pair of ions can be selected to detected a specific analyte, which is the so-called selected reaction monitoring (SRM) or multiple reaction monitoring (MRM) when more pairs are chosen in one GC-MS run. SRM and MRM of MS/MS are highly specific and virtually eliminate matrix background due to the two stages of mass selections. It can be applied to quantify trace levels of target compounds in the presence of sample matrices (Santos & Galceran, 2003). This has further secured its important role in modern discovery researches even in the era of fast liquid chromatography tandem mass spectrometry (LC-MS/MS) (Krone et al., 2010).

In a 3-D ion trap MS, the above-mentioned steps such as ion selection, ion dissociation and scanning of product ions occur in a timed sequence in a single trap in contrast to a triple quadrupole MS, where they are at different spatial locations of the instrument while ions travelling through. For a GC-MS/MS coupling, ion trap instruments offer three significant advantages over triple quadrupoles including low cost, easy to operate and high sensitivity for scanning MS/MS (Larsson & Saraf, 1997). Although in the earlier versions of ion trap instruments, a drawback called low-mass cut-off (which affects or discriminates small fragment ions in lower 1/3 mass range during MS/MS scan) has been observed, this has been solved in the modern instruments nowadays.

Although triple quadrupoles and ion traps have significant superiority in tandem mass spectrometry, their resolving powers are limited. Quadrupoles reach about unit resolution and ion trap a bit higher. With them, accurate mass measurements are generally not possible. This is just opposite to high-resolution TOF mass analysers. Another limitation of a triple quadrupole is that can be used only for one step of MS/MS, while an ion trap can perform multiple steps tandem MS experiments, which is sometimes very useful for structure elucidation.

2.1.4 GC with high resolution mass spectrometry (GC-TOF-MS)

Nowadays, the most investigated and applied coupling of GC with high resolution mass spectrometry is GC-TOF-MS, rather than that with sector instruments (Jeol, 2002). Recent advancements in instrumental optics design, the use of fast recording electronics and improvements in signal processing have led to a booming of the TOF-MS for investigation of organic compounds in complex matrices (Čajka & Hajšlová, 2007). GC-TOF-MS with high resolution of about 7000 is capable of achieving a mass accuracy as good as 5 ppm for small molecules. This allows not only isobaric ions to be easily mass-resolved but also the measurements of accurate masses for elemental composition assignment or mass confirmation, which adds one more powerful means for identification in GC-MS besides mass spectrum database searching and tandem mass spectrometry (Hernández et al, 2011). The moderate acquisition speed (maximum rate 10/s) at high resolution and the high acquisition speed (maximum rate of 500/s) at unit resolution can be achieved readily with a linear range of three to four orders of magnitude. High acquisition speed of GC-TOF-MS at unit resolution is very compatible with very narrow chromatographic peaks eluted from a fast and ultra-fast GC or GC×GC (Čajka & Hajšlová, 2007; Hernández et al, 2011; Pasikanti et al., 2008).

High resolution detection in GC-TOF-MS offers not only the high mass accuracy of molecular and fragment ions but also the accurate isotopic distribution with regard to isotope intensities and isotope-resolved information for element assignments. It is extremely helpful for unknown compounds for which no Library spectrum is available for database searching as reviewed recently (Hernández et al, 2011). With the help of software tools, a carbon number prediction filter can be applied to reduce the number of possible elemental compositions based on the relative abundance of the isotopic peak corresponding to ^{13}C (relative to the ^{12}C peak, each ^{13}C isotope contributes 1.1% to the ^{13}C peak). The nitrogen rule can also be used to determine whether the ion is an "even-electron ion" (for instance, protonated or deprotonated molecule) or an "odd-electron ion" (for instance, radical cation or anion) (MaLafferty, 1993). With these considerations, possible elemental compositions can be obtained when it is searched in available databases (e.g., Index Merck, Sigma Aldrich, Chem-Spider, Pubchem, Reaxys) and a chemical structure can be proposed (Hernández et al, 2011; Portolés et al., 2011). Both accurate masses and isotopes of fragment ions should be in agreement with the chemical structure assigned. However, in order to secure this identification, a reference standard will be required in a final step to check the GC retention time and to confirm the presence of fragment ions experimentally by GC-TOF-MS analysis (Hernández et al, 2011).

After the first applications of high resolution GC-TOF-MS such as on extracts of well water (Grange, Genicola & Sovocool, 2002), this approach has been well accepted and has triggered the rapid instrumentation and software developments. GC-TOF-MS has been proven very useful for target screening of organic pollutants in water, pesticide residues in food, anabolic steroids in human urine and xenoestrogens in human-breast tissues as well as non-target screening (Hernández et al, 2011). It has also been successfully applied in metabolomics. As an example, GC-TOF-MS coupled to an APCI source was applied to human cerebrospinal fluid for metabolic profiling (Carrasco-Pancorbo et al., 2009). Moreover, the use of GC×GC coupled to TOF-MS for the metabolic profiling of biological fluids has been discussed in a recent review (Pasikanti et al., 2008). Although it is a very helpful tool, studies have also indicated that the elucidation of unknowns cannot be achieved by following a standardized procedure, as both expertise and creativity are essential in the process (Portolés et al., 2011).

Despite its excellent mass accuracy and sensitivity for qualitative studies, GC-TOF-MS is not as robust as other MS detectors such as triple quadrupoles for quantitation due to its limited dynamic range.

2.1.5 GC coupled with tandem mass spectrometry

The use of capillary gas chromatography coupled with tandem mass spectrometry (GC-MS/MS) can in principle result in a superlative technique in terms of both sensitivity and specificity, as necessary for ultra trace analysis.

For qualitative identification with MS/MS, product ion scan, precursor ion scan and neutral loss with a triple quadrupole or product scan with an ion trap can be used. However, the high-resolution performance of a capillary GC also requires a MS/MS with high scanning speed. Its applications had been limited for a period until the recent decade when the scan speed of a quadrupole has been dramatically improved with an enhanced sensitivity instead of loss. This has greatly extended its applicability for trace unknown identification. For quantification with SRM or MRM, a short dwell time in order to reach sufficient data points for a narrow peak is also required. Improvements on this aspect has been from hundreds milliseconds reduced to a few millisecond for a triple quadrupole instrument. Typical applications are mainly found in contaminations in environment and foods such pesticides and PCBs in foods and biological samples (Krumwiede & Huebschmann, 2008). A large number of applications of tandem mass spectrometry in GC-ITMS for trace analysis can be found in the literature, for example, in environmental analysis (Santos & Galceran, 2003), microorganism characterisation (Larsson & Saraf, 1997) and forensic analysis (Chiarotti & Costamagna, 2000).

2.1.6 Two-dimensional GC coupled with mass spectrometry (GC×GC-MS)

Nowadays a one-dimensional GC offers a peak capacity in the range of 500-1000 (Mondello et al., 2008). Mass spectrometry already adds another dimension (resolving *m/z* of ions) to enhance GC-MS selectivity. However, it is not surprising that samples (e.g., extracts of natural products or metabolites) containing several hundreds or even thousands of volatile constituents are of common occurrence. With the increasing demands and improvements on sensitivity, many trace components have also to be investigated. The fact is that, in a one-dimensional GC separation, analytes are generally not equally distributed along the whole retention time scale but frequently co-elute. As a result, it is required that the system should have much higher peak capacity than the number of sample constituents in order to accommodate all sample components (Davis & Giddings, 1993; Mondello et al., 2008). It is obvious that comprehensive two-dimensional gas chromatography (GC×GC) can significantly enhance the resolving power of a separation (Liu & Phillips, 1991). Its developments and applications have been reviewed recently (Mondello et al., 2008). With thousands of compounds eluting at any time in both dimensions, MS detector has a superior power over any others with regard to sensitivity and specificity. Since the first application of GC×GC-MS (Frysinger & Gaines, 1999), its potential has been gradually exploited for analysis of petroleum, PCBs and food samples and complex extracts of natural products or metabolites (Dallüge et al., 2002; Pasikanti et al., 2008). The advantages of comprehensive two-dimensional gas chromatography in combination with mass spectrometry include unprecedented selectivity (three separation dimensions with regard to volatility, polarity, and mass), high sensitivity (through band compression), enhanced separation power, and increased speed (comparable to ultra-fast GC experiments, if the number of peaks resolved per unit of time is considered) (Mondello et al., 2008). With the development of interfacing techniques, fast MS scanning and data interpretation/presentation methods, it will become an important tool to uncover the wealth of undiscovered information with respect to the complex composition of samples.

2.2 Gas Chromatography-Fourier Transform Infrared Spectroscopy (GC-FTIR)

For novel structures or new chemical entities, it is possible that no matched reference spectra can be found in MS databases. Manual interpretation of mass spectra requires sound knowledge on organic mass spectrometry and dedicated experience and often not possible to suggest any candidate structures. With the help of the molecular spectroscopy FT-IR, information on functional groups or structure moieties with specific infrared absorptions is complementary to MS and can be very valuable for structure elucidation, as already being used as stand-alone.

After the inspiration of the first reported IR spectrum of a trapped GC peak in 1967 (Low & Freeman, 1967) and the first commercial instrument introduced in 1971 by Digilab, there has been a rapid development of hyphenated GC-FTIR techniques. Commonly for on-line GC-

FTIR coupling, the effluent from the GC flows through a heated transfer-line into the light-pipe. A schematic drawing of a typical GC-FTIR is shown in Figure 2. The interferograms are scanned continuously to record either 'on-the-fly' gas-phase vapour IR spectra or trapped component spectra (Jackson et al., 1993). This makes it possible to reconstruct a chromatogram in real time by a vector technique called the Gram-Schmidt method (Malissa, 1983). After the acquisition is finished, the spectra of each GC peak can be normalized and searched by comparison with an IR spectra library (Hanna et al., 1979; Seelemann, 1982). In the earlier days of this technique, there were many discussions and studies about its sensitivity followed by the possible overloading of the columns when a capillary GC is used. Nowadays a GC-FTIR system having *ng* sensitivity of absolute substance amount has been introduced (Bruker Optics, 2009). GC-FTIR has been applied for analysis of polychlrorinated dibenzo-p-dioxins, dibenzofurans (Sommer et al., 1997), aromatic polymers (Oguchi et al., 1991), petroleum (Scharma et al., 2009), etc., respectively. The combination of a gas chromatograph with both FT-IR and MS detectors on one instrument allows the simultaneous measurement of one peak by two supplementary detections. In fact, at each retention time, two different chromatograms were obtained. The sample passes the IR detector without destruction and is registered by the subsequent MS detector (Sommer et al., 1997).

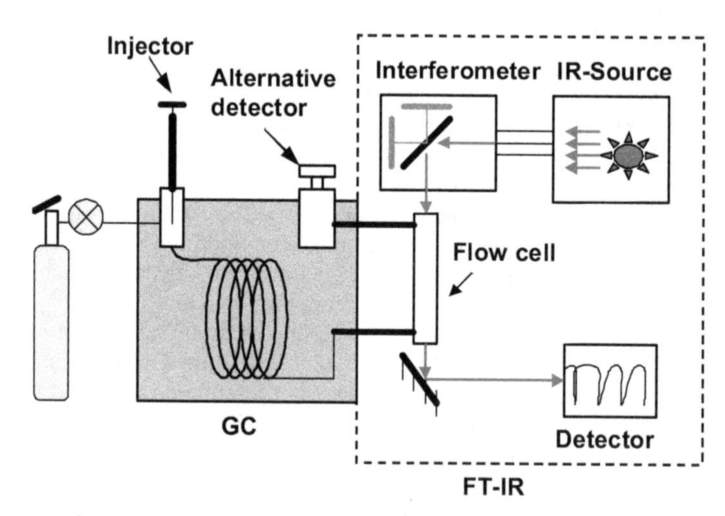

Fig. 2. Schematic drawing of a typical GC-FTIR

2.3 Gas Chromatography-Nuclear Magnetic Resonance spectroscopy (GC-NMR)

Nuclear Magnetic Resonance (NMR) (Rabi, 1938) spectroscopy is considered one of the most powerful analytical techniques for structural elucidation and identification of unknown compounds. It utilizes a physical phenomenon in which magnetic nuclei in a magnetic field absorb and re-emit electromagnetic radiation. Although the most commonly studied nuclei have specific chemical shifts, the abundances of the useful naturally occurring isotopes are generally low. That has significantly limited the sensitivity of NMR. Often mg of pure substance is required in order to obtain a meaningful spectrum. However, organic compounds of synthetic or natural products are often not in the pure form but are found as mixtures. Normally the analysis of such mixtures is performed in two independent steps: separation of the mixture into individual components and followed by identification. An on-line combination with GC separation can apparently accelerate this study.

The investigation on the coupling of GC and NMR has been very promising development because of the valuable information provided by NMR on molecular structure for each separated component (Buddrus & Herzog, 1981; Herzog & Buddrus, 1984). Unlike liquid or solid samples commonly analysed by NMR, the carrier gas in GC causes experimental difficulties in handling and results in low signal-to-noise ratio of the NMR signals obtained at atmospheric pressure. With the applications of Fourier-transform and averaging techniques, and gases can now be studied at fairly low pressures. However, the recording of the NMR spectra of flowing gases, especially those leaving a gas chromatograph containing only trace amount of analytes, imposes some problems. At short dwell times, the line broadening makes the recording of the spectra nearly impossible. Considerable progress towards overcoming its relative lack of sensitivity has been achieved using microcells for the optimum use of available samples and computers to improve signal-to-noise ratio (Milazzo, Petrakis & Brown, 1968). With the aid of stronger magnets and the newly developed microprobes, the first online GC-NMR spectra was recorded recently (Grynbaum et al., 2007; Kühnle et al., 2008). These experiments revealed the high potential of this technique, but it was also shown that the peaks in the GC separation elute up to 10 min, so that the advantage of the high separation performance of GC was lost in the experiment. The required amount for a GC-NMR run was between 1 and 2 mg for each analyte (Grynbaum et al., 2007). Problems arise in the form of partial condensation in the capillary connections (the transfer capillary and the probe head) for analytes with boiling points above 65 °C. It is suggested that the use of a transfer capillary being heated by a bifilar coil constructed from zero-susceptibility wire in combination with a strong magnetic field may solve overcome this problem (Grynbaum et al., 2007). Using stopped-flow measurements, very low sample amounts (~ 100-300 µg at 400 MHz), the potential applications of the hyphenation of high performance capillary GC to microprobe 1H NMR detection with the help of a spectra database has been demonstrated and an identification of stereoisomers in a complex mixture has been achieved (Kühnle et al., 2008). However, nanograms of volatile small compounds such as cockroach sex pheromones, mosquito attractants as well as a model compound geranyl acetate, which were prepared with an off-line GC, have already been investigated successfully using ^{1}H and COSY NMR (Nojima et al., 2011). Since the hyphenation of enantioselective capillary gas chromatography and mass spectrometry is not always sufficient to distinguish between structural isomers, thus requiring peak identification by NMR spectroscopy. The first online coupling of enantioselective capillary gas chromatography with proton nuclear magnetic resonance spectroscopy has been reported (Künhle et al. 2010). NMR allows constitutional and configurational isomers (diastereomers and enantiomers) to be distinguished. Enantiomers display identical spectra at different retention times, which enable an indirect identification of those unfunctionalized alkanes. Further developments in this field are still highly desirable.

2.4 Hyphenated gas chromatographic techniques for elemental analysis

Hyphenated techniques involving ICP-MS, AAS and AES are among the fastest growing research and application areas in atomic spectroscopy. The preferred chromatographic separation techniques include GC, HPLC, capillary electrophoresis (CE) and ion chromatography (IC) as well as field flow fractionation (FFF). General procedure for speciation analysis using GC-hyphenation include separation of the analytes from the sample matrix, formation of volatile derivatives, pre-concentration / cleanup and determination using different elemental analysis techniques.

2.4.1 Gas Chromatography - Inductively Coupled Plasma - Mass Spectrometry (GC-ICP-MS)

Stand-alone ICP-MS does not provide information on the chemical structure of the analyte at molecular level since all forms of the analytes are converted to positively charged atomic ions in the plasma. However, as an excellent elemental analyser (ICP) with resolution on masses (MS), ICP-MS can also be used as gas chromatographic detector. Resulting from this hyphenation, target analytes are separated into their constituent chemical forms or oxidation states before elemental analysis. In GC-ICP-MS, where the sample is gaseous, the transfer line should be inactivated and heated to eliminate sample degradation and condensation and will guide the sample directly into the ICP torch. In this way, the sample is maintained at constant high temperature from the end of the chromatographic column in the GC oven to the tip of the ICP injector (Agilent, 2007). It is almost a universal detector (only H, He, Ag, F, Ne cannot be directly measured), fits perfectly with a wide range of GC carrier gases and flows, and is capable of quantification with isotope dilution. A *pg* level of sensitivity can be achieved. The combination of the high performance chromatographic separation and the ICP atomizer also eliminates some interference of co-eluting compounds. For example in organic analysis, it has been used to study trace halogenated solvent residues in edible oils (Gomez-Ariza & Garcia-Barrera, 2006). GC-ICP-MS is very useful technique for speciation analysis such as sulphur speciation (Gomez-Ariza & Garcia-Barrera, 2006) or organometallic speciation (Bouyssiere, 2001; Kresimon et al., 2001).

2.4.2 Gas Chromatography-Atomic Absorption Spectroscopy (GC-AAS) and -Atomic Emission Spectroscopy (GC-AES)

Atomic absorption spectrometry (AAS) is a well-established detection technique in elemental analysis. In AAS, when the UV light with the right wavelength irradiates on a ground-state free atom, the atom may absorb the light to enter an excited state in a process known as atomic absorption. The wavelength of the light is somehow specific for an element. By measuring the amount of light absorbed, a quantitative determination of the amount of analyte element present can be made. A careful selection of wavelengths allows the specific quantitative determination of individual elements in the presence of others. By supplying enough thermal energy (such as in a flame), analyte compounds dissociate into free atoms. The ease and speed, at which precise and accurate determinations can be achieved, have made AAS one of the most popular methods for the analysis of metals.

Atomic emission spectroscopy (AES), also called optical emission spectroscopy (OES), is another elemental analysis technique in atomic spectroscopy. In this process, analytes are first atomized using either ICP or microwave irradiation (high temperatures), where the atoms are raised to electronically excited states. As they return to lower electronic energy levels, photons are emitted at wavelengths that are characteristic of the particular element. Both the wavelength and intensity of the emitted photons will be recorded. This data may be used to identify the elements present and quantify them after a calibration. GC-AES determines the elemental composition for each GC peak.

The tailored operation condition of a gas chromatograph fits again nicely with the AAS atomizer since the analytes are already present in the gas phase. The coupling can be done by introducing the GC effluent via a heated nickel transfer line into the quartz atomization furnance (Dirkx et al., 1992). Although the interface is rather simple, in practice, several conditions need to be optimized in order to obtain good sensitivity and selectivity for specific analytical problems. Efforts include exploiting the quartz furnace, heating with flame or a thermostat, or using the graphite furnace as the atomization device (Dirkx et al., 1995). The widespread use of organometallic compounds in agriculture, plastics industry and subsequently the release into the environment has led to an increasing concern about their persistence and toxicity. For instance, organotins unlike inorganic tin, it may occur at toxic levels in aquatic and sedimentary environments. The distribution of organotins as well as its bioavailability and toxicity depend critically on the chemical form in which the species are actually present. This is why total tin determination is not able to provide reliable information on the hazards of this element to human health and environment. Therefore, speciation analysis of these organometallic compounds are of great importance (Dirkx et al., 1995). Since its introduction, GC-AAS has been extensively applied to speciation of organometallic compounds such as Sn, Pb and As. Typical examples include speciation of organolead in environment (Baxter & Frech, 1995; Dirkx et al., 1992; Harrison & Hewitt, 1985) and in biological samples (Baxter & Frech, 1995); speciation of organotin compounds in agriculture and plastics industry (Dirkx et al., 1995) as well as arsenic speciation. The reason for the existence of various forms of organometallic compounds is that Sn, Pd and As all have very rich organic chemistry. Just because of this, several derivatization methods have been reported in order to convert them to volatile species for GC analysis and a sensitivity of picogram level can be achieved together with a highly selective detection as the result of the coupling (Dirkx et al., 1995; Harrison & Hewitt, 1985).

As for GC-AES, it has been much less investigated than GC-AAS. Nevertheless, the available reports have indicated that GC-AES coupled with microwave-induced plasma (MIP) can be used to study organic polymers (Oguchi et al., 1991) and to perform speciation analysis of organotin compounds in human urine. This provides a method suitable for rapid sensitive screening of human urine samples without dilution of the sample (Zachariadis & Rosenberg, 2009a, 2009b).

3. Conclusion and remarks

Apparently, the motivation of the enormous efforts on investigating a variety of hyphenated systems is the increasing demands on the wealth of undiscovered information of trace components in a sample. Many stand-alone instrumental analytical techniques are already well developed and able to provide specific information about analytes in samples, although each of them has its advantages and disadvantages. The complexity of real-world sample matrices often exceeds the analytical capabilities of any conventional chromatographic separations. The development and employment of more comprehensive coupling techniques to enable a deeper insight into the composition of natural and synthetic matrices has become a necessity.

The most powerful separation technique in chromatography is still capillary gas-liquid chromatography (one-dimensional or its extension - comprehensive two-dimensional gas chromatography). On the other hand, mass spectrometry including various high resolution MS and tandem MS provides analyte-specific selectivity. Their combination GC-MS is still the most flexible and powerful analytical technique available today for volatile organic substances with many unchallengeable advantages such as unprecedented selectivity, sensitivity, enhanced separation power and speed of analysis (Mondello et al., 2008).

The coupling with other detection techniques further extends the GC applicability. However, the rule of thumb is that the hyphenation of two techniques would only make sense when the analytical powers (separation and detection) of both instruments are enhanced rather than compromised in order to couple. One of the recent good examples is implementation of the very soft ionization APCI interface for GC-TOF-MS (Carrasco-Pancorbo et al., 2009) to generate abundant molecular ions in order to utilize the high resolution and MS/MS capabilities for structure elucidation. Both CI and EI may be too hard to preserve valuable molecular ions. This feature makes this technique very attractive for such as wide-scope screening of a large number of analytes at trace levels. In the future developments, efforts should be put to make hyphenated GC techniques also user-friendly and robust with regard to automatic data interpretation and reporting software, standardized interfaces and even consumables.

4. References

Agilent Technologies. (2007). *Handbook of Hyphenated ICP-MS Applications*, 1st ed. Part No. 5989-6160EN, August 2007

Allcin, C.; Law, R. & Morris, S. (1999). Polybrominated Diphenylethers in Sediments and Biota Downstream of Potential Sources in the UK. *Environmental Pollution*, Vol. 105, No. 2 (May 1999), pp. 197-207, ISSN 0269-7491

Arsenault, G.; Dolhun, J. & Biemann, K. (1971). Alternate or Simultaneous Electron Impact-Chemical Ionization Mass Spectrometry of Gas Chromatographic Effluent. *Analytical Chemistry*, Vol. 43, No. 12 (October 1971), pp. 1720-1722, ISSN 0003-2700

Baxter, D. & Frech, W. (1995). Speciation of Lead in Environmental and Biological Samples. *Pure and Applied Chemistry*, Vol. 67, No. 4 (April 1995), pp. 615-648, ISSN 0033-4545

Bouyssiere, B.; Szpunar, J. & Lobinski, R. (2001). Gas Chromatography with Inductively Coupled Plasma Mass Spectrometric Detection in Speciation Analysis. *Spectrochimica Acta Part B*, Vol. 57, No. 5 (May 2002), pp. 805–828, ISSN 0584-8547

Bruker Optics. (2009). GC-FT-IR in the *ng* Range. Application Note # AN-76E. Available from:
http://www.brukeroptics.com/fileadmin/be_user/Applications/ApplicationNotes/AN_76E_GC-FT-IR_in_the_ng_range.pdf

Buddrus, J. & Herzog, H. (1981). Coupling of Chromatography and NMR. 3. Study of Flowing Gas Chromatographic Fractions by Proton Magnetic Resonance. *Organic Magnetic Resonance*, Vol. 15, No. 2 (February 1981), pp. 211-213, ISSN 1097-458X

Čajka, T. & Hajšlová, J. (2007). Gas Chromatography–Time-of-Flight Mass Spectrometry in Food Analysis. *LCGC Europe*, Vol. 20, No. 1 (2007), pp. 25–31, ISSN 1471-6577

Carrasco-Pancorbo, A.; Nevedomskaya, E.; Arthen-Engeland, T.; Zey, T.; Zurek, G.; Baessmann, C.; Deelder, A. & Mayboroda, O. (2009). Gas Chromatography

/Atmospheric Pressure Chemical Ionization-Time of Flight Mass Spectrometry: Analytical Validation and Applicability to Metabolic Profiling. *Analytical Chemistry,* Vol. 81, No. 24 (December 2009), pp. 10071-10079, ISSN 0003-2700

Chaturvedi, K. & Nanda, R. (2010). Hyphenated Gas Chromatography. *International Journal of Pharmaceutical Sciences Review and Research,* Vol. 5, No. 3 (November 2010), pp. 18-27, ISSN 0976-044x

Chernetsova, E.; Revelsky, A.; Revelsky, I.; Mikhasenko, I & Sobolevsky, T. (2002). Determination of Polychlorinated Dibenzo-p-Dioxins, Dibenzofurans, and Biphenyls by Gas Chromatography / Mass Spectrometry in the Negative Chemical ionization Mode with Different Reagent Gases. *Mass Spectrometry Review,* Vol. 21, No. 6 (November 2002), pp. 373– 387, ISSN 1098-2787

Chernushevich, I.; Loboda, A. & Thomson, B. (2001). An Introduction to Quadrupole–Time-of-Flight Mass Spectrometry. *Journal of Mass Spectrometry,* Vol. 36, No. 8, (August 2001), pp. 849–865, ISSN 1096-9888

Chiarotti, M. & Costamagna, L. (2000). Analysis of 11-Nor-9-carboxy-Δ^9-tetrahydrocannabinol in Biological Samples by Gas Chromatography Tandem Mass Spectrometry (GC/MS-MS). *Forensic Science International,* Vol. 114, No. 1 (October 2000), pp. 1-6, ISSN 0379-0738

Cody, R. (2008). GC/MS with DART Ion Source. Application Note, JEOL Inc., 06.09.2011. Available from http://chromatographyonline.findanalytichem.com/lcgc/data/articlestandard//lcgc/122008/504148/article.pdf

Cody, R.; Laramée J. & Durst H. (2005). Versatile New Ion Source for the Analysis of Materials in Open Air under Ambient Conditions. *Analytical Chemistry,* Vol. 77, No. 8 (March 2005), pp. 2297-2302, ISSN 0003-2700

Coleman III, W. & Gordon, B. (2006). Hyphenated Gas Chromatography, In: *Encyclopedia of Analytical Chemistry,* R.A. Meyers, (Ed.), John Wiley & Sons, ISBN: 9780470027318

Cotter, R. (1994). *Time-of-Flight Mass Spectrometry.* American Chemical Society, ISBN 0-8412-3474-4 , Columbus, USA

Dallüge, J. ; van Stee, L. ; Xu, X. ; Williams, J. ; Beens, J.; Vreuls, R. & Brinkman, U. (2002). Unravelling the Composition of Very Complex Samples by Comprehensive Gas Chromatography Coupled to Time-of-Flight Mass Spectrometry: Cigarette Smoke. *Journal of Chromatography A,* Vol. 974, No. 1 (October 2002), pp. 169–184, ISSN 0021-9673

Davis, J. & Giddings, J. (1983). Statistical Theory of Component Overlap in Multicomponent Chromatograms. *Analytical Chemistry,* Vol. 55, No. 3 (March 1983), pp. 418-424, ISSN 0003-2700

de Hoffmann, E. & Stroobant, V. (2003). *Mass Spectrometry: Principles and Applications (Second Ed.).* John Wiley & Sons, ISBN 0471485667, Toronto, Canada

Dirkx, W.; van Cleuvenbergen, R & Adams, F. (1992). Speciation of Alkyllead Compounds by GC-AAS: A State of Affairs. *Microchimica Acta,* Vol. 109, No. 1-4 (1992), pp. 133-135, ISSN 0026-3672

Dirkx, W.; Lobinski, R. & Adams, F. (1995). Speciation Analysis of Organotin by GC-AAS and GC-AES after Extraction and Derivatization, In: *Quality Assurance for Environmental Analysis*, Ph. Quevauviller, E.A. Maier and B. Griepink, (Ed.), 357-409, Elsevier, ISSN 0-444-89955-3, Amsterdam, The Netherlands

Friedrich, J. & Acree, T. (1998). Gas Chromatography Olfactometry (GC/O) of Dairy Products. *International Dairy Journal*, Vol. 8, No. 3 (March 1998), pp. 235-241, ISSN 0958-6946

Frysinger, G. & Gaines, R. (1999). Comprehensive Two-Dimensional Gas Chromatography with Mass Spectrometric Detection (GC×GC/MS) Applied to the Analysis of Petroleum. *Journal of High Resolution Chromatography*, Vol. 22, No. 5, (May 1999), pp. 251-255, ISSN 0935-6304

Gohlke, R. & McLafferty, F. (1993). Early Gas Chromatography/Mass Spectrometry. *Journal of the American Society for Mass Spectrometry*, Vol. 4, No. 5 (May 1993), pp. 367-371, ISSN 1044-0305

Gomez-Ariza, J. & Garcia-Barrera, T. (2006). Optimization of a Multiple Headspace SPME-GC-ECD-ICP-MS Coupling for Halogenated Solvent Residues in Edible Oils. *Journal of Analytical Atomic Spectrometry*, Vol. 21, No. 9 (2006), pp. 884-890, ISSN 0267-9477

Goraczko, A. (2005). Molecular Mass and Location of the Most Abundant Peak of the Molecular Ion Isotopomeric Cluster. *Journal of Molecular Modeling*, Vol. 11, No. 4 (April 2005), pp. 271–277, ISSN 1610-2940

Grange, A.; Genicola, F. & Sovocool, G. (2002). Utility of Three Types of Mass Spectrometers for Determining Elemental Compositions of Ions Formed from Chromatographically Separated Compounds. *Rapid Communications in Mass Spectrometry*, Vol. 16, No. 24, (December 2002), pp. 2356–2369, ISSN 1097-0231

Grange, A.; Winnik, W.; Ferguson, P. & Sovocool, G. (2005). Using a Triple-Quadrupole Mass Spectrometer in Accurate Mass Mode and an Ion Correlation Program to Identify Compounds. *Rapid Communications in Mass Spectrometry*, Vol. 19, No. 18 (September 2005), pp. 2699–2715, ISSN 1097-0231

Gross, M. (1994). Accurate Masses for Structure Confirmation. *Journal of the American Society for Mass Spectrometry*, Vol. 5, No. 2 (February 1994), pp. 57-57, ISSN 1044-0305

Grynbaum, M.; Kreidler, D.; Rehbein, J.; Purea, A.; Schuler, P.; Schaal, W.; Czesla, H.; Webb, A.; Schurig, V. & Albert, K. (2007). Online Coupling of Gas Chromatography to Nuclear Magnetic Resonance Spectroscopy: Method for the Analysis of Volatile Stereoisomers. *Analytical Chemistry*, Vol. 79, No. 7 (April 2007), pp. 2708-2713, ISSN 0003-2700

Grob, R. & Barry, E. (2004). *Modern Practice of Gas Chromatography*, Wiley-Interscience, ISBN 0-471-22983-0, New York, USA

Hanna, A.; Marshall, J. & Isenhour, T . (1979). A GC/FT-IR Compound Identification System. *Journal of Chromatographic Science*, Vol. 17, No. 8 (August 1979), pp. 434-, ISSN 0021-9665

Harrison, A. (1992). *Chemical Ionization Mass Spectrometry*, CRC Press, ISBN 0849342546, Boca Raton, USA

Harrison, R. & Hewitt, C. (1985). Development of Sensitive GC-AAS Instrumentation for Analysis of Organometallic Species in the Environment. *International Journal of Environmental Analytical Chemistry*, Vol. 21, No. 1-2 (December 1985), pp. 89-104, ISSN 0306-7319

Hernández, F.; Portolés, T.; Pitarch, E. & López, F. (2011). Gas Chromatography Coupled to High-Resolution Time-of-Flight Mass Spectrometry to Analyze Trace-Level

Organic Compounds in the Environment, Food Safety and Toxicology. *Trends in Analytical Chemistry*, Vol. 30, No. 2 (February 2011), pp. 388-400, ISSN 0165-9936

Herzog, H & Buddrus, J. (1984). Coupling of Chromatography and NMR. Part 5: Analysis of High-Boiling Gas-Chromatographic Fractions by On-line Nuclear Magnetic Resonance. *Chromatographia*, Vol. 18, No. 1, pp. 31-33, ISSN 0009-5893

Hirschfeld, T. (1980). The Hy-phen-ated Methods. *Analytical Chemistry*, Vol. 52, No. 2 (February 1980), pp. 297A-312A, ISSN 0003-2700

Holmes, J. & Morrell F. (1957). Oscillographic Mass Spectrometric Monitoring of Gas Chromatography. *Applied Spectroscopy*, Vol. 11, No. 2 (May 1957), pp. 61-91, ISSN 0003-7028

Hunt, D.; Stafford, Jr. G.; Crow, F. & Russel, J. (1976). Pulsed Positive Negative Ion Chemical Ionization Mass Spectrometry. *Analytical Chemistry*, Vol. 48, No. 14 (December 1976), pp. 2098-2104, ISSN 0003-2700

Jackson, P.; Dent, G.; Carter, D.; Schofield, D.; Chalmers, J.; Visser, T. & Vredenbregt, M. (1993). Investigation of High Sensitivity GC-FTIR as an Analytical Tool for Structural Identification. *Journal of High Resolution Chromatography*, Vol. 16, No. (September 1993), pp. 515-521, ISSN 0935-6304

Jaeger, H. (1987). *Capillary Gas Chromatography-Mass Spectrometry in Medicine and Pharmacology*, Dr. Alfred Huethig, ISBN 3-7785-1375-3, Heidelberg, Germany

James, A. & Martin, J. (1952). Gas-Liquid Partition Chromatography; the Separation and Micro-estimation of Volatile Fatty Acids from Formic Acid to Dodecanoic Acid. *Biochemistry Journal*, Vol. 50, No. 5 (March 1952), pp. 679-690, ISSN 0264-6021

Jeol. (2002). Analysis of Scotch Whiskey and Tequila Samples by Solid-Phase Microextraction and High-Resolution GC/MS. Application Note: MS-11262002-A, 15.09.2011. Available from: http://www.jeolusa.com/PRODUCTS/AnalyticalInstruments/MassSpectrometers/GCmateII/tabid/231/Default.aspx

Karasek, F. & Clement, R. (1988). *Basic Gas Chromatography – Mass Spectrometry Principles & Techniques*. Elsevier, ISBN 0-444-42760-0, Amsterdam, the Netherlands

Kingdon, K. (1923). A Method for the Neutralization of Electron Space Charge by Positive Ionization at Very Low Gas Pressures. *Physical Review*, Vol. 21, No. 4 (April 1923), pp. 408-418, ISSN 0031-899X

Krone, N.; Hughes, B.; Lavery, G.; Stewart, P.; Arlt, W. & Shackleton, C. (2010). Gas Chromatography/Mass Spectrometry (GC/MS) Remains a Pre-Eminent Discovery Tool in Clinical Steroid Investigations Even in the Era of Fast Liquid Chromatography Tandem Mass Spectrometry (LC/MS/MS). *Journal of Steroid Biochemistry & Molecular Biology*, Vol. 121, No. 3 (August 2010), pp. 496–504, ISSN 0960-0760

Krumwiede, D. & Huebschmann, H. (2008). Analysis of PCBs in Food and Biological Samples Using GC Triple Quadrupole GC-MS/MS. Application Note 10262, Thermo Scientific. Available from: http://pops.thermo-bremen.com/pdf /AN10262_PCBs_on_TSQ_Quantum_GC.pdf

Kuhara, T. (2005). Gas Chromatographic-Mass Spectrometric Urinary Metabolome Analysis to Study Mutations of Inborn Errors of Metabolism. *Mass Spectrometry Review*, Vol. 24, No. 6 (November 2005), pp. 814-827, ISSN 1098-2787

Kühnle, M.; Kreidler, D.; Holtin, K.; Czesla, H.; Schuler, P.; Schaal, W.; Schurig, V. & Albert, K. (2008). Online Coupling of Gas Chromatography to Nuclear Magnetic Resonance Spectroscopy: Method for the Analysis of Volatile Stereoisomers. *Analytical Chemistry*, Vol. 80, No. 14 (July 2008), pp. 5481-5486, ISSN 0003-2700

Kühnle, M.; Kreidler, D.; Holtin, K., Czesla, H., Schuler, P., Schurig, V. & Albert, K. (2010). Online Coupling of Enantioselective Capillary Gas Chromatography with Proton Nuclear Magnetic Resonance Spectroscopy. *Chirality*, Vol. 22, No. 9 (October 2010), pp. 808-812, ISSN 1520-636X

Liu, Z. & Phillips, J. (1991). Comprehensive Two-Dimensional Gas Chromatography using an On-column Thermal Desorption Modulator Interface, *Journal of Chromatographic Science*, Vol. 29, No. 6, (June 1991), pp. 227–231, ISSN 0021-9665

Larsson, L. & Saraf, A. (1997). Use of Gas Chromatography-Ion Trap Tandem Mass Spectrometry for the Detection and Characterization of Microorganisms in Complex Samples. *Molecular Biotechnology*, Vol. 7, No. 3 (June 1997), pp. 279-287, ISSN 1559-0305

Low, M. & Freeman, S. (1967). Measurement of Infrared Spectra of Gas-Liquid Chromatography Fractions using Multiple-Scan Interference Spectrometry. *Analytical Chemistry*, Vol. 39, No. 2 (February 1967), pp. 194-198, ISSN 0003-2700

Makarov A. (2000). Electrostatic Axially Harmonic Orbital Trapping: A High-Performance Technique of Mass Analysis. *Analytical Chemistry*, Vol. 72, No.6 (February 2000), pp. 1152-1162, ISSN 0003-2700

Malissa, H. (1983). On the Use of Capillary Separation Columns in GC/FTIR-Spectroscopy and on the Quantitative Evaluation of the Gram-Schmidt Reconstructed Chromatogram. *Fresenius' Journal of Analytical Chemistry*, Vol. 316, No. 7 (January 1983), pp. 699-704, ISSN 0937-0633

Mamyrin, B.; Karataev, V.; Shmikk, D. & Zagulin, V. (1973). The Mass-Reflectron, a New Nonmagnetic Time-of-Flight Mass Spectrometer with High Resolution. *Soviet Physics JETP*. Vol. 37, (July 1973), pp. 45, ISSN 0038-5646

Mamyrin, B. (2001). Time-of-Flight Mass Spectrometry (Concepts, Achievements, and Prospects)". *International Journal of Mass Spectrometry*, Vol. 206, No. 3 (March 2001), pp. 251–266, ISSN 1387-3806

Marshall, A.; Hendrickson, C. & Jackson, G. (1998). Fourier Transform Ion Cyclotron Resonance Mass Spectrometry: a Primer. *Mass Spectrometry Review*, Vol. 17, No. 1 (January 1998), pp. 1-35, ISSN 1098-2787

McLafferty, F.; Hertel, R. & Villwock, R. (1974). Probability Based Matching of Mass Spectra. Rapid Identification of Specific Compounds in Mixtures. *Organic Mass Spectrometry*, Vol. 9, No. 7 (July 1974), pp. 690–702, ISSN 1096-9888

McLafferty, F. & Tureek, F. (1993). *Interpretation of Mass Spectra*. 4th ed.. University Science Books, ISBN 0935702253, Sausalito, USA

McNaught, A & Wilkinson, A. (1997). *IUPAC. Compendium of Chemical Terminology*, 2nd ed. (the "Gold Book"). Blackwell Scientific Publications, ISBN 0-9678550-9-8, Oxford, UK.

Message, G. (1984). *Practical Aspects of Gas Chromatography / Mass Spectrometry*, Wiley, ISBN 0-471-06277-4, New York, USA

Milazzo, B.; Petrakis, L. & Brown, P. (1968). Microsampling Techniques for Combined Gas Chromatography and High-Resolution Nuclear Magnetic Resonance Spectroscopy, *Applied Spectroscopy*, Vol. 22, No. 5 (September 1968), pp. 574-575, ISSN 0003-7028

Mondello, L.; Tranchida, P.; Dugo, P & Dugo, G. (2008). Comprehensive Two-Dimentional Gas Chromatography-Mass Spectrometry: A Review. *Mass Spectrometry Review*, Vol. 27, No. 2 (March 2008), pp. 101-124, ISSN 1098-2787

Niessen, W. (2001). *Current Practice of gas chromatography-mass spectrometry*, Marcel Dekker, ISBN 0-8247-0473-8, New York, USA

Nojima, S; Kiemle, D.; Webster, F.; Apperson, C & Schal, C. (2011). Nanogram-Scale Preparation and NMR Analysis for Mass-Limited Small Volatile Compounds. *Public Library of Science (PLoS) One*, Vol 6, No. 3 (March 2011), e18178, eISSN -1932-6203

Oguchi, R.; Shimizu, A.; Yamashita, S.; Yamaguchi, K & Wylie, P. (1991). Polymer Analysis Using Pyrolysis-GC-FTIR-MS and GC-AED. *Journal of High Resolution Chromatography*, Vol. 14, No. 6 (June 1991), pp. 412–416, ISSN 0935-6304

Ong, V. & Hites, R. (1994). Electron Capture Mass Spectrometry of Organic Environmental Contaminants. *Mass Spectrometry Review*, Vol. 13, No. 3 (May 1994), pp. 259-283, ISSN 1098-2787

Pasikanti K.; Ho, P. & Chan, E. (2008). Gas Chromatography/Mass Spectrometry in Metabolic Profiling of Biological Fluids. *Journal of Chromatography B*, Vol. 871, No. 2 (August 2008), pp. 202–211, ISSN 1570-0232

Paul, W. & Steinwedel, H. (1953). Ein Neues Massenspektrometer ohne Magnetfeld. *Zeitschrift für Naturforschung A*, Vol. 8, No. 7, pp. 448-450, ISSN 0932-0784

Peterson, A.; McAlister, G.; Quarmby, S.; Griep-Raming, J. & Coon, J. (2010). Development and Characterization of a GC-Enabled QLT-Orbitrap for High-Resolution and High-Mass Accuracy GC/MS. *Analytical Chemistry*, Vol. 82, No. 20 (October 2010), pp. 8618-8628, ISSN 0003-2700

Plomley, J; Lausevic, M. & March, R. (2000). Determination of Dioxins/Furans and PCBs by Quadrupole Ion-Trap Gas Chromatography–Mass Spectrometry. *Mass Spectrometry Review*, Vol. 19, No. 5 (September 2000), pp. 305-365, ISSN 1098-2787

Portolés, T.; Pitarch, E.; López, F.; Hernández, F. & Niessen, W. (2011). Use of Soft and Hard Ionization Techniques for Elucidation of Unknown Compounds by Gas Chromatography / Time-of-Flight Mass Spectrometry. *Rapid Communications in Mass Spectrometry*, Vol. (25), No. 11 (June 2011), pp. 1589–1599, ISSN 1097-0231

Rabi, I.; Zacharias, J.; Millman, S. & Kusch, P. (1938). A New Method of Measuring Nuclear Magnetic Moment. *Physical Review*, Vol. 53, No.4 (April 1938), pp. 318-318, ISSN 0031-899X

Santos, F. & Galceran, M. (2003). Modern Developments in Gas Chromatography-Mass Spectrometry-based Environmental Analysis. *Journal of Chromatography A*, Vol. 1000, No. 1 (June 2003), pp. 125-151, ISSN 0021-9673

Sarrion, M; Santos, F. & Galceran, M. (2000). In Situ Derivatization/Solid-Phase Microextraction for the Determination of Haloacetic Acids in Water. *Analytical Chemistry*, Vol. 72, No. 20 (October 2000), pp. 4865-4873, ISSN 0003-2700

Schwartz, J., Senko, M. & Syka, J. (2002). A Two-Dimensional Quadrupole Ion Trap Mass Spectrometer. *Journal of the American Society for Mass Spectrometry*, Vol. 13, No.6 (June 2002), pp. 659-669, ISSN 1044-0305

Seelemann, R. (1982). GC/FTIR Coupling - a modern tool in analytical chemistry. *Trends in Analytical Chemistry*, Vol. 1, No. 14 (October 1982), pp. 333-339, ISSN 0003-2700

Sharma, K; Sharmab, S. & Lahiri, S. (2009). Characterization and Identification of Petroleum Hydrocarbons and Biomarkers by GC-FTIR and GC-MS. *Petroleum Science and Technology*, Vol. 27, No. 11 (June 2009), pp. 1209 – 1226, ISSN 1091-6466

Solouki, T.; Szulejko, J.; Bennett, J. & Graham, L. (2004). A Preconcentrator Coupled to a GC/FTMS. Advantages of Self-Chemical Ionization, Mass Measurement Accuracy, and High Mass Resolving Power for GC Applications. *Journal of the American Society for Mass Spectrometry*, Vol. 15, No. 8 (August 2004), pp. 1191-1200, ISSN 1044-0305

Sommer, S.; Kamps, R.; Schumm, S. & Kleinermanns, K. (1997). GC/FT-IR/MS Spectroscopy of Native Polychlorinated Dibenzo-p-dioxins and Dibenzofurans Extracted from Municipal Fly-Ash. *Analytical Chemistry*, Vol. 69, No. 6 (March 1997), pp. 1113-1118, ISSN 0003-2700

Sparkman, D. (2006). *Mass Spec Desk Reference*, 2nd. ed., Global View Publishing, ISBN 0966081390, Pittsburgh, USA.

Stephens, W. (1946). A Pulsed Mass Spectrometer with Time Dispersion. *Physical Review*, Vol. 69, No. 11 (June 1946), pp. 691-691, ISSN 0031-899X

Stein, S. & Scott, D. (1994). Optimization and Testing of Mass Spectral Library Search Algorithms for Compound Identification. *Journal of the American Society for Mass Spectrometry*, Vol. 5, No. 9 (Sptember 1994), pp. 859-866, ISSN 1044-0305

Szulejko, J. & Solouki, T. (2002). Potential Analytical Applications of Interfacing a GC to an FT-ICR MS: Fingerprinting Complex Sample Matrixes. *Analytical Chemistry*, Vol. 74, No.14 (July 2002), pp. 3434-3442, ISSN 0003-2700

Todd, J. (1991). Recommendations for Nomenclature and Symbolism for Mass Spectroscopy (Including an Appendix of Terms Used in Vacuum Technology). *Pure and Applied Chemistry*, Vol. 63, No. 10 (October 1991), pp. 1541-1566, ISSN 0033-4545

Vestal, M. & Campbell, J. (2005). Tandem Time-of-Flight Mass Spectrometry. *Methods in Enzymology*, Vol. 402, pp. 79–108, ISSN 978-0-12-182807-3

Watson, J. (1997). *Introduction to Mass Spectrometry*, Lippincott-Raven, ISBN 0397516886, Philadelphia, USA

Yergey, J.; Heller, D.; Hansen, G.; Cotter, R. & Fenselau, C. (1983). Isotopic Distributions in Mass Spectra of Large Molecules. *Analytical Chemistry*, Vol. 55, No. 2 (February 1983), pp. 353–356, ISSN 0003-2700

Zachariadis, G. & Rosenberg, E. (2009a). Speciation of Organotin Compounds in Urine by GC-MIPAED and GC-MS after Ethylation and Liquid-Liquid Extraction. *Journal of Chromatography B*, Vol. 877, No. 11-12 (April 2009), pp. 1140-1144, ISSN 1570-0232

Zachariadis, G. & Rosenberg, E. (2009b). Determination of Butyl- and Phenyltin Compounds in Human Urine by HS-SPME after Derivatization with Tetraethylborate and Subsequent Determination by Capillary GC with Microwave-Induced Plasma Atomic Emission and Mass Spectrometric Detection. *Talanta*, Vol. 78, No. 2 (April 2009), pp. 570-576, ISSN 0039-9140

2

Design, Modeling, Microfabrication and Characterization of the Micro Gas Chromatography Columns

J.H. Sun, D.F. Cui, H.Y. Cai,
X. Chen, L.L. Zhang and H. Li
State Key Laboratory of Transducer Technology, Institute of Electronics, Chinese Academy of Sciences, Beijing, China

1. Introduction

Gas chromatography (GC) systems can separate different components of gaseous mixtures, and are important analytical tools for a variety of disciplines, including environmental analysis, methane gas probes, and homeland security and pollution monitoring. However, most modern GC devices suffered the problems of low detection speed, sensitivity, and poor stability. In addition, these conventional GC devices were bulky and fragile, which ruled out the possibility of the in-field use. In some cases, gas samples were collected in the field and analyzed in the laboratory using a conventional GC, which was inconvenient and inefficient.

Combined with micro-detector (such as micro thermal conductivity detector, micro photoionization detector, etc.), Micro GC columns can be developed into a miniaturized chromatographic system because of its very small size. This integrated GC system, with a small size, light weight, rapid analysis, high sensitivity, easy to use, etc., can be widely applied to environmental pollution, home safety, pesticide residues, food safety, pre-diagnosis of cancer and other areas for achieving on-site and on-line rapid testing.

In this chapter, several micro gas chromatography columns were designed for building micro μGC systems, and this chapter will contain the following sections: Structural design, modeling analysis, microfabrication and characterization of the micro gas chromatography columns.

2. Structural consideration

Structure of the μGC columns can be designed in accordance with the requirements of the designer. Several various structural designs are appeared in research papers, but their differences in-depth analysis of these designs are almost absence. In fact, structure is an important factor for affecting the separation performance of the μGC columns, because the shape of the column, especially the mutations sections (such as the corner of the channel),will change the airflow velocity, pressure distribution, the thickness of the stationary phase film and other important factors, moreover, these factors are key factors for affecting the separation performance of μGC columns.

In this work, two frequently-used configurations (the spiral channel and the serpentine channel, as shown in Fig.1) of columns were designed for GC analysis, the shape of cross-section is rectangular, and their sizes are consistent. Then, the effect of airflow rate and pressure distribution in the channel was simulated using ANSYS analysis. Especially, the effect of these factors in the corners was the focus.

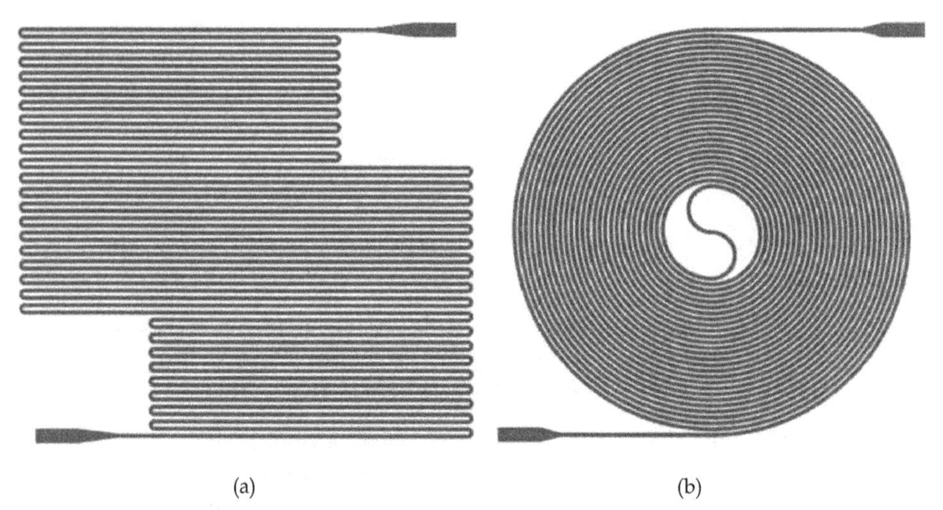

(a) (b)

Fig. 1. The structure design, (a) the serpentine channel, (b) the spiral channel, for μGC column.

3. Theoretical consideration and simulation

3.1 Theoretical consideration

If the volume and the concentration of a sample are small enough (about μmol) and in the linear range of adsorption isotherm, the elution curve equation can be determined by equation (1) based on the plate theory.

$$C = \frac{\sqrt{n}}{\sqrt{2\pi}} \bullet \frac{m}{V_r} \bullet \exp[-\frac{n}{2}(1-\frac{V}{V_r})^2] \tag{1}$$

Where C is the sample concentration at any point of the elution curve, m is the weight of the solute, V is the retention volume at any point of the elution curve, V_r is the retention volume of the solute, n is the number of theoretical plates. When $V = V_r$, C reaches its max value:

$$C_{\max} = \frac{\sqrt{n} \bullet m}{\sqrt{2\pi} \bullet V_r} \tag{2}$$

Derived from the elution curve equation, the number of theoretical plates n can be defined by the following equation:

$$n = 5.54\left(\frac{t_r}{w_{1/2}}\right)^2 \tag{3}$$

Where t_r is the retention time, and $w_{1/2}$ is the width of the peak at half height.

And the theoretical plate height (H) can be determined by:

$$H = \frac{L}{n} \tag{4}$$

Where L is the length of the column. According to the equation (3) and (4), as the chromatography peak's bottom width (w) decreases, the number of theoretical plates (n) increases and the theoretical plate heigh (H) decreases accordingly, which results in higher column efficiency. Hence, n and H are the index of column efficiency.

Due to some effect factors for H can't be shown in equation (4), the theoretical plate height (H) can also be given by the following formula:

$$H = \frac{2D_g}{\bar{u}} f_1 f_2 + \frac{(1 + 9k + 25.5k^2)}{105(k+1)^2} \frac{w^2 \bar{u}}{D_g} \frac{f_1}{f_2} + \frac{2}{3} \frac{k}{(k+1)^2} \frac{(w+h)^2 d_f^2}{D_s h^2} \bar{u} \tag{5}$$

where D_g and D_s are the binary diffusion coefficients in the mobile and stationary phases, respectively, d_f is the thickness of the stationary phase, w and h are the channel width and height, respectively, k is the retention factor, and f_1 and f_2 are the Giddings–Golay and Martin–James gas compression coefficients, respectively. The average linear airflow velocity is given by

$$\bar{u} = \frac{w^2 p_0 (P^2 - 1)}{24\eta L} f_2 \tag{6}$$

Where P_0 is the outlet pressure, P is the ratio of the inlet to outlet pressure, L is the column length, and η is the carrier gas viscosity.

From the equation (5) and (6), the airflow rate and pressure are key factors for the separation performance of the μGC columns.

Resolution is called overall separation efficiency, which is defined as the difference of retention time between two adjacent chromatography peaks divided by the half of the sum of these two peak's bottom width:

$$R = \frac{t_{r_2} - t_{r_1}}{\frac{1}{2}(w_1 + w_2)} = \frac{2(t_{r_2} - t_{r_1})}{w_1 + w_2} \tag{7}$$

The definition of resolution (R) in equ (7) does not reflect all the factors which influence resolution, because resolution is actually determined by column efficiency (n), selectivity factor (α) and capacity factor (k), hence the resolution also can be described by equation (8):

$$R = \frac{\sqrt{n}}{4}\left(\frac{\alpha - 1}{\alpha}\right)\left(\frac{k}{1+k}\right) \tag{8}$$

After the stationary phase is chosen, the selectivity factor could be correspondingly fixed, which means that the resolution is only affected by n. For a column with a certain theoretical plate height, square of the resolution is proportional to the length of column:

$$\left(\frac{R_1}{R_2}\right)^2 = \frac{n_1}{n_2} = \frac{L_1}{L_2} \tag{9}$$

Therefore the major approach to improve resolution is to increase the column length.

3.2 Simulation

Base on chromatography theory, the airflow rate and pressure are key factors for the separation performance of the μGC columns. An obviously different variation of the airflow velocity, pressure distribution and other factors will be appeared in these μGC columns due to their different structural, and these variations will affect the separation performance of μGC columns. So a detailed comparative analysis for the two frequently-used configurations of columns is made in this paper, which provides a theoretical basis for designing ideal μGC columns.

In this chapter, the effects on the airflow rate and pressure were simulated using ANSYS analysis. Because the major effect of airflow rate is come from the mutation section (such as the corner of GC-channel) of these channels, the corner of the chromatographic channel are selected for the simulation. In the simulation, the first step is modelling. The parameters of the modelling are consistent with the size of the actual channel (The cross-section is rectangular, the width of the channel is 150 μm, and the depth of the channel is 100 μm).

After modelling, boundary conditions were set as follows: the airflow rate in the gas inlet was set to 18 cm/s, the pressure of the gas outlet was set to 0, the displacement of the other part was set to 0).

Fig.2 (a) and (b) show the solutions of the simulation, the airflow rate in the corner of serpentine channel is 15.73 cm/s, the difference of airflow rate between in the corners of serpentine channel and in the gas inlet reaches 3.13 cm/s. However, the airflow rate in the mutation section of the spiral channel is 16.93 cm/s, the difference of airflow rate between in the corners of spiral channel and in the gas inlet is only 1.33 cm/s. Moreover, the changes in distribution of airflow rate in the corners of the serpentine channel are obvious and relatively large. But the distribution of airflow rate in the corners of the spiral channel is relatively uniform. Fig.3 (a) and (b) show the solutions of the pressure distribution in the serpentine channel and spiral channel, the simulation results show that there exists a pressure gradient in these two kinds of channel, but the change rate of pressure in serpentine channel is obviously much larger than that of the spiral channel, the maximum pressure value in spiral channel is only 32.80 Pa, however, the maximum pressure value in the serpentine channel is over 150.0 Pa, the value is close to 5 times compared to the former. Seen from the above analysis, the impact on airflow rate and pressure in serpentine channel is very significant, and these effects will lead to deterioration of separation performance. Because the difference of the airflow rate in the channel would change the thickness of the stationary phase film. Consequently, the variation of the thickness of the stationary phase film in the channel would lead bad tailing peaks.

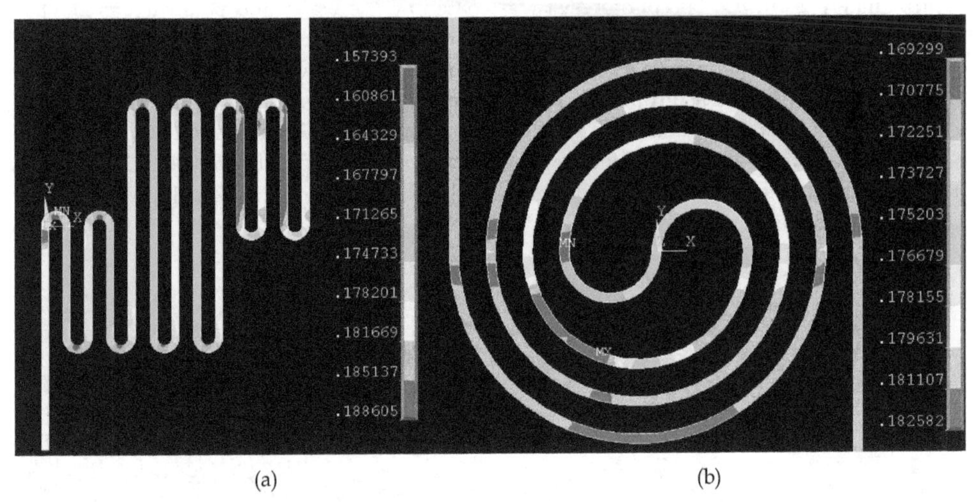

Fig. 2. The effect of the airflow rate in the corners of the channel (a)the serpentine channel,(b) the spiral channel, were simulated using ANSYS analysis.

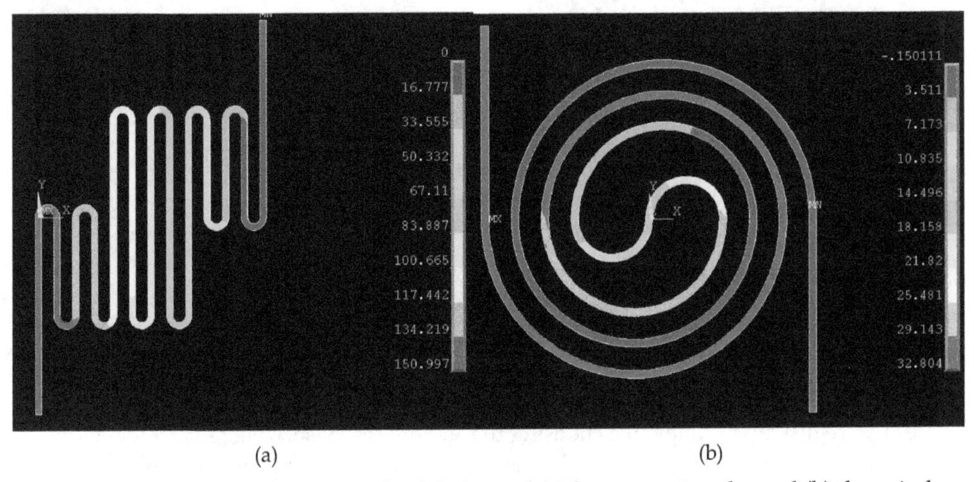

Fig. 3. The pressure distribution in the GC channel (a)the serpentine channel,(b) the spiral channel, were simulated using ANSYS analysis.

4. Microfabrication

4.1 Column fabrication

Fabrication of the μGC column includes aluminium deposition, photolithography and deep reactive-ion etching (DRIE) (the fabrication process is illustrated in Fig.4). Firstly, a 2-μm-thick electron-beam evaporation aluminium film was deposited on a p-type <100> silicon wafer which served as the etch mask in following steps. Secondly, a thickness of approximately 2 μm AZ1500 photoresist was coated on the wafer and patterned as an etch mask for aluminium. Subsequently, aluminium without the protection of photoresist was etched away by an etchant (H_3PO_4) and the silicon surface was exposed. Then, a DRIE process, instead of the anisotropically KOH chemical etching process, was utilized to form the rectangular micro channels. Finally, all aluminium masks were removed and the silicon wafer and pyrex7740 glass were held together and heated to approximately 400℃. A 1000 V potential was then applied between the glass and the silicon, and the resulting electrostatic force pull the wafers into intimate contact.

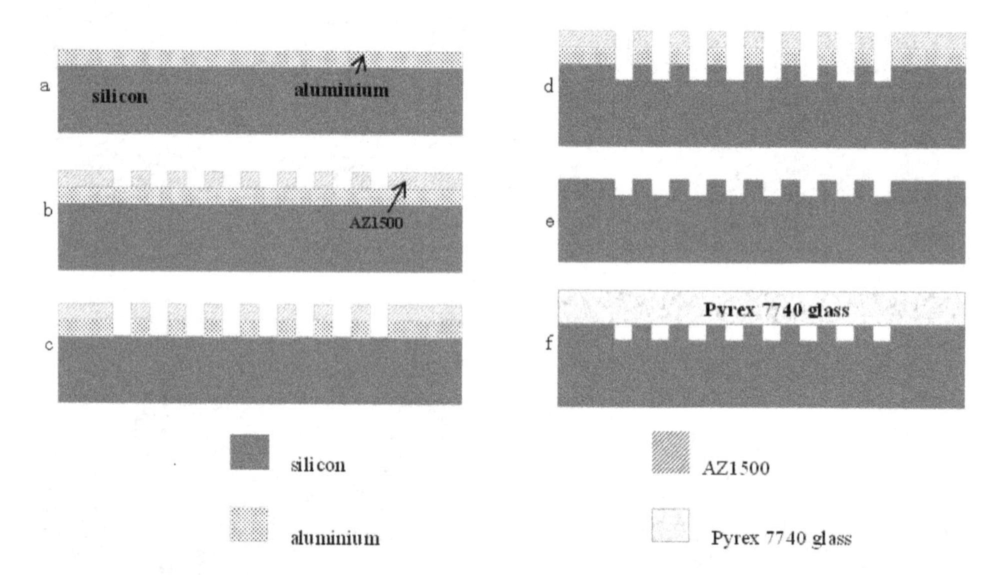

Fig. 4. The GC column process flow.

4.2 The stationary phase coating

The uniformity, stability and thickness of the stationary phase film are important factors that can affect the separation efficiency. In this section, the selection principle of stationary phase, coating methods and coating procedure are introduced.

There is no regularity to follow for choosing the stationary phase. Generally, "Like dissolves like" principle has been identified as the basic theory for selecting the stationary phase. In the application, the selection of stationary phase should be determined according to the actual situation.

i. Separation of non-polar compounds: Generally, non-polar stationary phase is used. Each component flows over the GC column according to the order of boiling point. The component with a low boiling point is the first one out of the GC column, and followed by the component with a high boiling point.
ii. Separation of polar compounds: Generally, polar stationary phase is used. The component with a small polar is the first one out of the GC column, and followed by the component with a relatively large polar.
iii. Separating non-polar and polar compounds: Generally, polar stationary phase is used. The non-polar component is the first one out of the GC column, and followed by the polar component.
iv. Separation of complex and difficult compounds: Two or more mixed stationary phase can be used.

Stationary phase coating methods are generally including static and dynamic coating. Static coating procedure is defined as: The stationary phase is filled with the GC column, one end is sealed, and the other end is connected a vacuum pump, the solvent is slowly evaporated under the pressure of vacuum pump, until all of the solvent is completely evaporated, and a stationary phase film with a thickness of 0.1-0.2 microns is left on the channel.

Dynamic coating procedure is defined as: A stationary phase solvent is injected into the GC column, and the stationary phase solvent flows through the GC column under pressure, the thickness of Stationary phase can be controlled by changing the flow rate and the concentration of the stationary phase solvent. The stationary phase solvent is pushed out from the other side of the GC column, and the nitrogen gas was delivered through the column for several hours to completely evaporate the solvent, and a stationary phase film with a thickness of 0.1-0.2 microns is left on the channel.

In this work, separation of benzene and homologue of benzene was taken as for example. In order to completely separate the mixture, OV-1 or OV-101 was the optimal stationary phase. So OV-1 was selected to coat the micro GC column using a dynamic coating procedure. The process was shown as following: Firstly, about 5 μl OV-1 was dissolved in 2.0 ml mixtures of n-pentane and dichloromethane with a volume ratio of 1:1. The mixture was agitated for 30 minutes to ensure full dissolution. Secondly, the inlet of the column was connected with a capillary, and the outlet was connected with a laboratory made micro-pump, which was used to inject the stationary phase solution into the GC column. After the GC column was full of the solution, the micro-pump was turned off for 30 minutes to make sure that it was long enough for the stationary phase to attach to the channel wall. Then, nitrogen gas was delivered through the column for a few hours to completely evaporate the n-Pentane and dichloromethane. Subsequently, the column was put into an oven under a nitrogen atmosphere in which the temperature of the oven was firstly increased gradually by $5\,^{\circ}C\,/\,min$ until $100\,^{\circ}C$ and then the temperature of the oven was kept at $100\,^{\circ}C$ for 4 hours. Fig. 5 shows an SEM view of the stationary phase film coated on the column wall, in which the film is uniform and adheres well to the channel. The composition analysis of the film (the selected area of the film on the channel wall can be seen from Fig. 6(a)), by X- Ray Photoelectron Spectroscopy, is described in the Fig. 6(b) which shows that the composition agrees with the composition of OV-1.

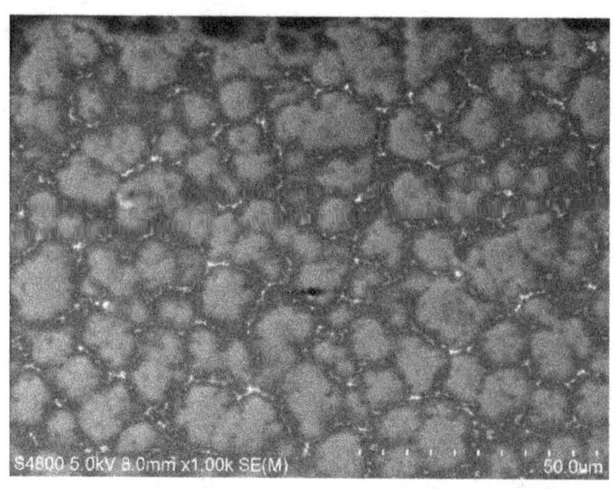

Fig. 5. The SEM view of the stationary phase film.

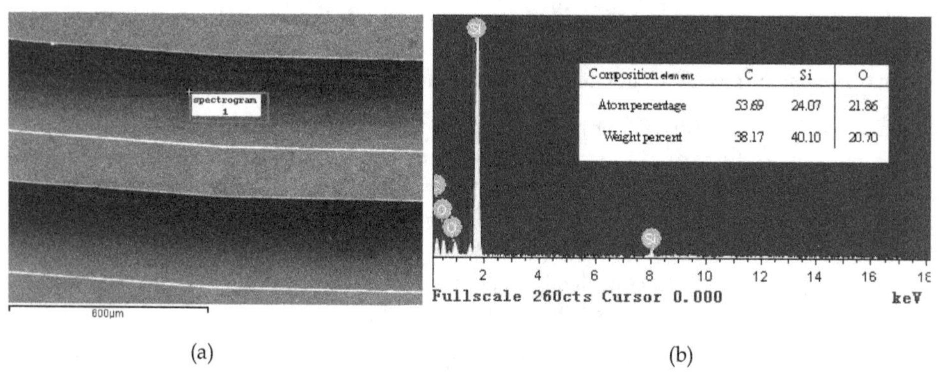

(a) (b)

Fig. 6. The composition analysis of the film which operated by X- Ray Photoelectron Spectroscopy, (a) the selected area of the film on the channel wall, (b) the percent of the composition element in the film.

4.3 Interface technology

The inlet/outlet interface technology of MEMS-based GC columns is a key factor. If not properly tackled, some problems, such as instability of the interface, air leakage from the interface and the airflow jamming would happen, which usually led to the failure of the GC columns. In order to solve these problems, a fixed base (the base dimensions is 8.0 mm long, 4.0 mm wide and 5.0 mm high, See Fig.7) was used to connect the tubes with the inlet /outlet of the column. The major steps are briefly covered: Firstly, a film of heat-resistant adhesive was coated on the bottom surface of the base, and then the base was aligned with the inlet or the outlet carefully and bonded with the column; After 30 minutes, more heat-resistant adhesive was coated around the base with a steel tube (Tubing, steel, $1/32'' \times 0.25mm\ ID$) into the base, and then some heat-resistant adhesives were coated around the joint points. Subsequently, a burst pressure test with 0.4 MPa pressure was applied on the column and the interface remained undamaged. Fig. 8 shows a photograph of the completed GC columns including the connected base.

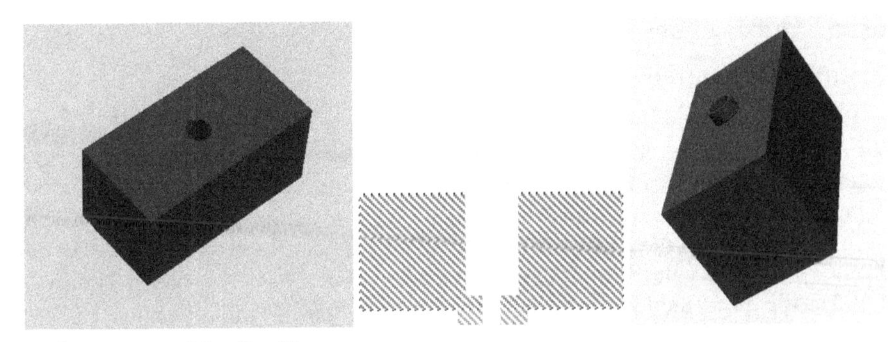

Fig. 7. The picture of the fixed base.

Fig. 8. The photograph of the completed GC columns, (a) 0.75m serpentine channel column, (b) 0.75m spiral channel column, and the chips dimension are all 2.2cm×2.2cm, (c) 3 m spiral channel column and the chip dimension is 3.5cm × 3.5, (d) SEM micrograph of the section and the channel.

5. Results and discussions

5.1 Experimental setup and test condition

GC performances were measured through the flame ionization detector (FID) method on an Agilent 7890A GC system. In the test, the carrier gas was helium and the original operating temperature of columns was 40℃, in order to reduce the total analysis time of the component mix, temperature programming, which the temperature is increased from 40°C to 80°C at a rate of 5°C /min, was used in the process of separating the sample. These samples were dissolved in 1.0 ml CS_2, and the volume of each component sample, which was injected at time zero, was 1 μl. The top-air method was used with a split ratio of 1:10.

5.2 Experimental results

In this section, the effect of the structure the flow rate and the length for separation performance were analyzed, and then the experiments for separating complex mixtures were studied.

5.2.1 Effect of the structure for separation performance

These columns were operated at the velocity of 18 cm/s, the sample is a 3-component mixture including benzene, toluene, o-xylene, and the chromatogram graphs were shown in Fig.9 (a) and (b). In Fig. 9 (a), the serpentine channel GC columns separated the gaseous mixture with a resolution of 8.9 (calculated by equation (7)) between toluene and o-xylene and yielded about 1900 theoretical plates (calculated by equation (3)). However, the chromatographic peak emerged bad tailing peaks. The inconsistent of the film thickness were the primary cause for the tailing peaks. However, the spiral channel GC columns yielded about 3900 theoretical plates and separated the gaseous mixture with the resolution of 10.97 (as shown in Fig. 9 (b)). Moreover, the chromatographic peaks were greatly improved. So the structure of GC channel, especially the design of the corner, was a key factor for the separation performance. The spiral channel possessed streamlined shape. As a result, the airflow couldn't be baffled, and the separation experiment also verified that the spiral channel GC column showed excellent overall separation performance and separation efficiency for the gas mixture. Thus, the spiral channel column is superior to the serpentine channel column for GC analysis. Therefore, on this basis, we designed a 3 m spiral column for GC application, then, the effect of the flow rate and the length for separation performance was analyzed and its separation performance for complex mixture was studied.

5.2.2 Effect of the flow rates for separation performance

Because the flow rate was a key factor for separation performance, different flow rates were used in the process of separating the sample (the sample is a 3-component mixture including benzene, toluene, o-xylene). Firstly, the 3 m spiral column was operated at the velocity of 42 cm/s, and the chromatogram graph was shown in Fig.10 (a). From experimental data, the number of plates can be calculated as given in equation (3). The 3 m spiral column yielded 3410 plates. The resolution, R, can be calculated by equation (7).The resolution between benzene and toluene was 4.05, and the resolution between toluene and o-xylene was 9.27. While, if the flow rate was appropriately reduced and the column was operated at the velocity of 18 cm/s, the column yielded approximately 6160 plates (the experimental data was shown in Fig.10 (b)), moreover, the resolution between benzene and toluene was improved from 4.05 to 8.02, and the resolution between toluene and o-xylene was improved from 9.27 to 14.3. The separation efficiency of the gas mixture was strongly improved.

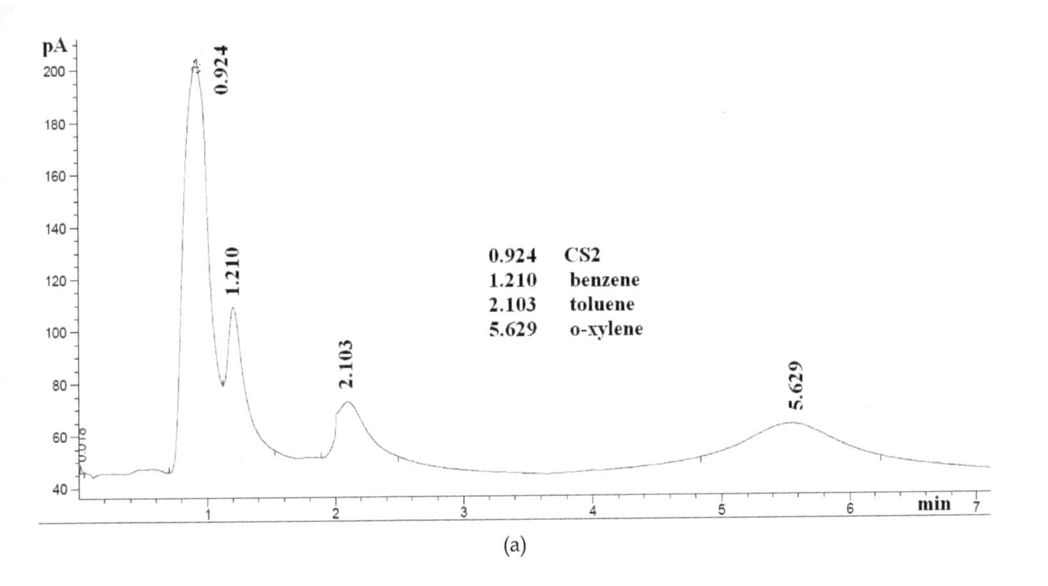

(a)

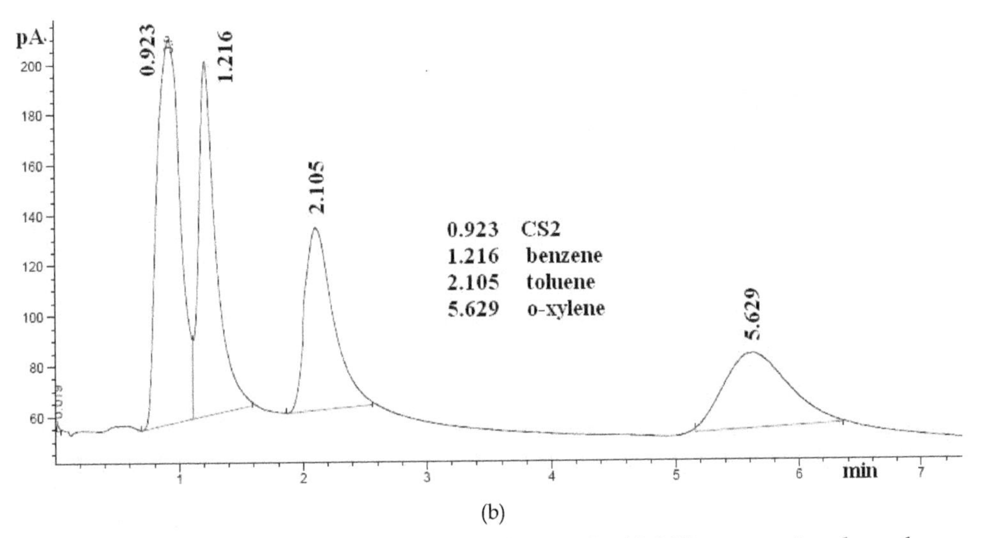

(b)

Fig. 9. Chromatogram of a gas mixture achieved using the (a) 0.75m serpentine channel column, (b) 0.75m spiral channel column.

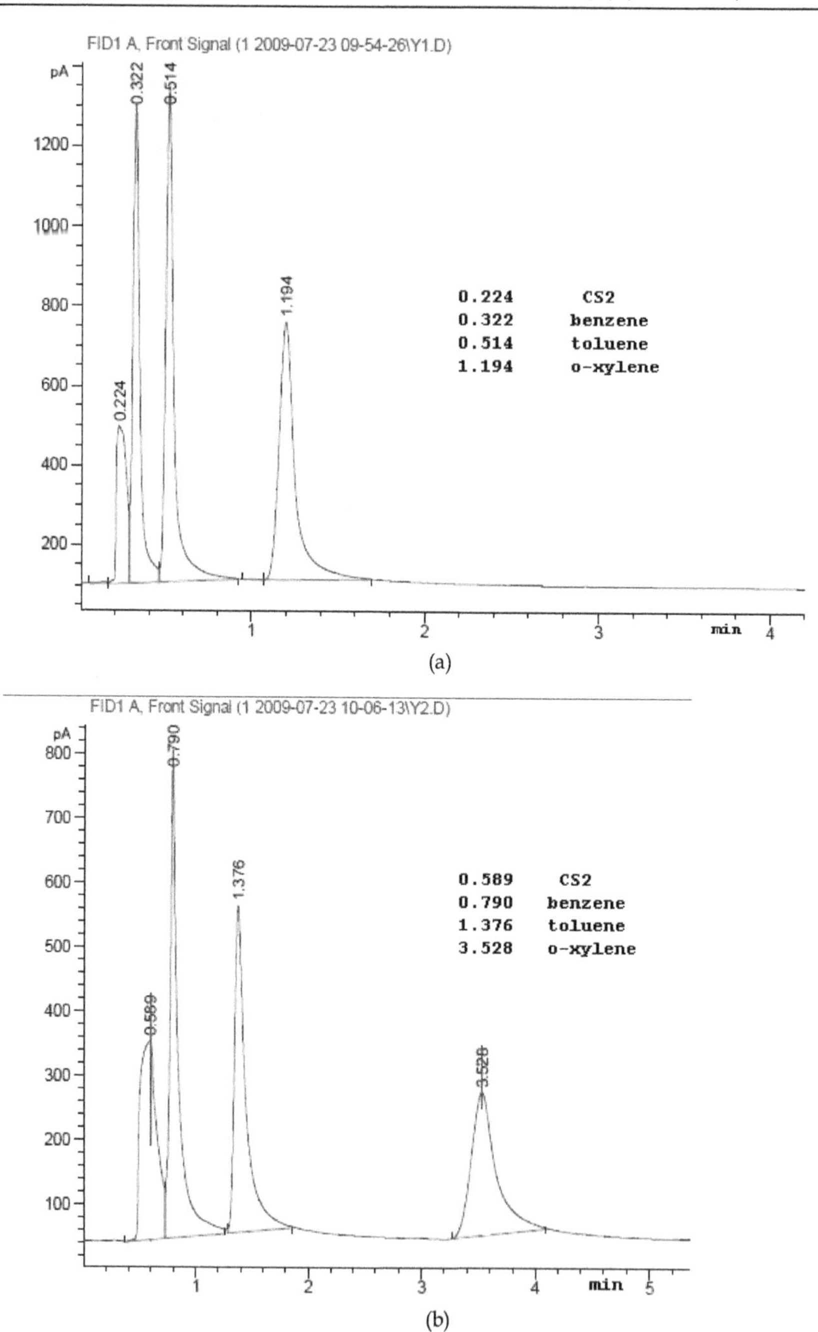

Fig. 10. Chromatogram of a gas mixture containing benzene, toluene and o-xylene achieved using the 3m column coated with OV-1, (a) operated at the flow rate of 42 cm/s, (b) operated at the optimal flow rate of 18 cm/s.

5.2.3 The effect of the column length for separation performance

Based on the formula (9), the major approach to improve resolution is to increase the column length.

So in this paper, the separation performance of GC columns with different length (such as the length of the column is 0.5 m, 1 m and 3 m, but the section width and depth of the column are the same) were compared with, and the separation experiments were all operated at the velocity of 18 cm/s and used the same sample (the sample including 3-component mixture: benzene, toluene, o-xylene). The columns yielded approximately 2400, 3370 and 6160 theoretical plates (the actual test values), respectively.

The theoretical predictions values could be also roughly computed 2500,5000 and 15000 plates(derived from the equation 1 and equation 4), respectively, The actual test values were about 41% of theoretical prediction values When the column length increased to 3 meters (Assuming the maximum retention factors is less than 10, the diffusion coefficients in the gas and liquid phases are about 0.1~0.2 cm^2/m, and the thickness of the stationary phase film is 0.2 μm), The comparison curve of the number of theoretical plates between the theoretical prediction values and the actual values was shown in Fig.11.

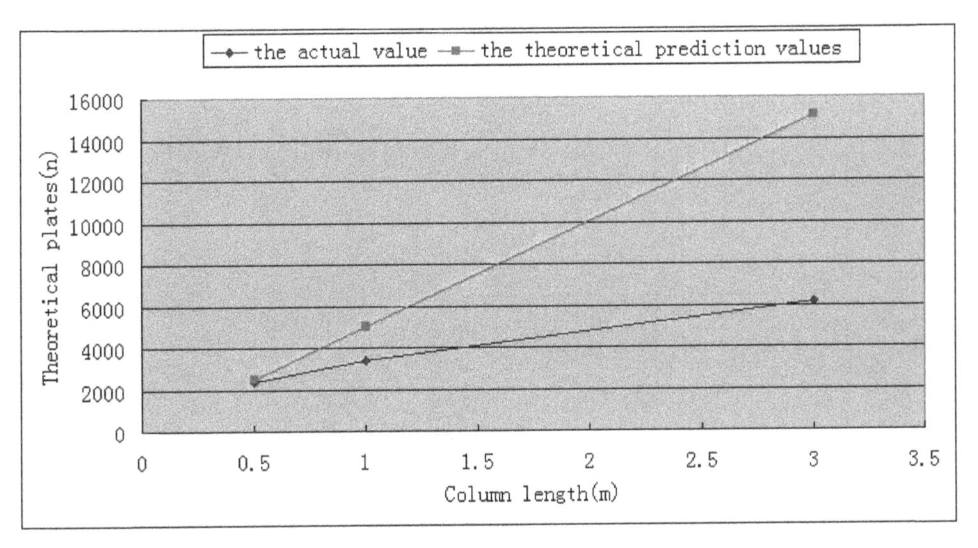

Fig. 11. The comparison curve of the number of theoretical plates between the theoretical prediction values and the actual values.

The discrepancy with the predicted number of the theoretical plates can be attributed to stationary phase film nonuniformity with the increase in column length. Consequently, the theoretical plates could be significantly increased by lengthening the GC column. Moreover, the separation performance of GC column can be greatly improved. The GC column can acquire a good overall separation performance when the length of the GC column is increased to 3 m.

5.2.4 Separation of the complex mixtures using 3 m spiral column

A separation experiment was performed to separate the sample mixture of benzene, toluene, ethylbenzene, p-xylylene and o-xylene using the microfabrication GC column. The column was operated at the velocity of 18 cm/s, and the chromatogram graph was shown in Fig.12. From experimental data, the spiral 3 m column yielded 7100 theoretical plates, and the analysis time was less than 200 sec. The GC column acquired a good overall separation performance.

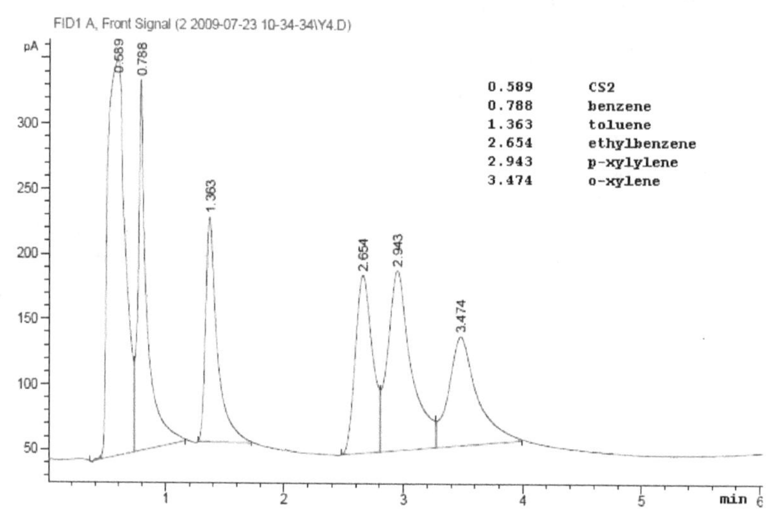

Fig. 12. Chromatogram of a gas mixture containing benzene, toluene, ethylbenzene, p-xylylene and o-xylene, achieved using the 3m column coated with OV-1 which operated at the flow rate of 18 cm/s.

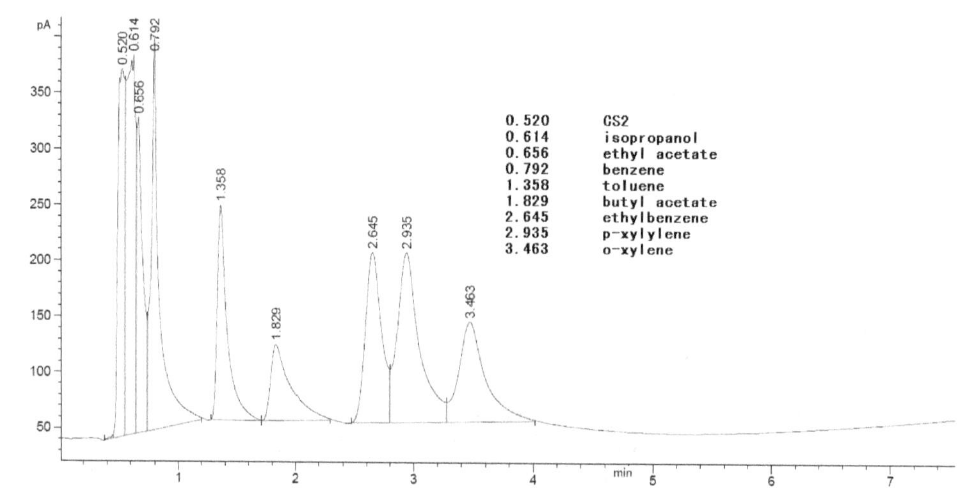

Fig. 13. Chromatogram of a gas mixture containing benzene, toluene, ethylbenzene, p-xylylene, o-xylene, butyl acetate, isopropanol and ethyl acetate achieved using the 3m column coated with OV-1 which operated at the flow rate of 18 cm/s.

Then, three other components were mixture into the sample, of which two kinds of components were polar components (isopropanol and ethyl acetate), and another sample was the butyl acetate. The separation experiment was carried out under the same conditions, the chromatograms was shown in the Fig.13, these polar components and solvents came out almost simultaneously, which resulted in these two polar components couldn't be separated by non-polar stationary phase, but the acid butyl ester was perfectly separated. The result indicated the non-polar column with high selectivity and specificity.

6. Conclusion

In summary, this chapter has presented the simulation, fabrication and experimental results for the microfabricated GC columns. Prototypes of devices have been fabricated successfully and the stationary phase is tackled properly. From the result of the simulation, the changes in distribution of airflow rate in the corners of the serpentine channel are obvious and relatively large, and the change rate of pressure in serpentine channel is obviously larger than that of the spiral channel, the value is close to 5 times compared to the former. Seen from the experimental data, the impact on airflow rate and pressure in serpentine channel is very significant, and this effect leads to deterioration of separation performance. Moreover, the difference of the airflow rate in the channel changes the thickness of the stationary phase film when the stationary phase is deposited. Consequently, the variation of the thickness of the stationary phase film in the channel leads bad tailing peaks. So, the spiral channel column is superior to the serpentine channel column for GC analysis. A series of spiral GC columns are designed according to the above analysis. The 3 m yields approximately 7100 plates and perfectly separates the complex mixture in less than 200 sec. Combined with micro-detector and micro sampler, Micro GC columns can be developed a miniaturized chromatographic system and serve as a platform technology for gas mixture separations.

Currently, the developed micro-columns can not achieve its own heating, but the design and fabrication process of the integrated micro-heater GC columns have been completed, its performance will be reported in subsequent papers. Micro- thermal conductivity detector, micro-optical ionization detector and micro-meter will be developed in our further work. A new generation of micro-chromatographic system will be developed by integrating micro GC column with the micro detectors. The micro GC system will be applied to environmental pollution, home safety, pesticide residues, food safety, pre-diagnosis of cancer and other areas for achieving on-site rapid testing.

7. Acknowledgment

The authors greatly acknowledge the financial support from the National Science Foundation of China under Grant number 61176112, 60976088 and 60701019. This work is sponsored by the Major National Scientific Research Plan (2011CB933202). The authors would like to thank Z.W.Ning and Y.N. Zhang and L.D.Du for their technical supports.

8. References

A. Bhushan, D. Yemane, D.Trudell, et al., (2007). Fabrication of micro-gas chromatograph columns for fast chromatography, Microsyst Technol 2007, Vol. 13, pp.361-368.

A.D. Radadia, R.I. Masel, M.A. Shannon, J.P. Jerrell, and K.R. Cadwallader (2008).Micromachined GC columns for fast separation of organophosphonate and organosulfur compounds,Analytical Chemistry, Vol. 80, No. 11, pp. 4087–4094.

G. R. Lambertus, C. S. Fix, S. M. Reidy, R. A. Miller, D. Wheeler, E. Nazarov, R. Sacks (2005),Silicon Microfabricated Column with Microfabricated Differential Mobility Spectrometerfor GC Analysis of Volatile Organic Compounds, Anal. Chem ,Vol.77,pp.7560-7571.

Golay, M. J. E. (1958), Gas Chromatography1958, D. Desty, Ed. Butterworths Sci. Pub., London, pp. 36–55.

J.H Sun, D.F.Cui, H.Y. Cai, H.Li, X. Chen, L.D. Du and L.L Zhang,Design, Simulation and fabrication of High Performance Gas Chromatography Columns for Analysis of Volatile Organic Compounds, Sensor Letters,Vol.9,No.2,pp.655-658. ISSN 1546-198X.

J.H Sun, D.F.Cui, H.Y. Cai, X. Chen, H. Li and L.D. Du(2010), Simulation and Evaluation of the Silicon-Micromachined Columns for Gas Chromatography, Chinese journal of chemistry,Vol. 28,pp.2315-2317. ISSN 1001-604X.

J.H.Sun, D.F.Cui, Y.T Li, L.L Zhang, J. Chen, H. Li and X.Chen,(2009).A High Resolution MEMS-based Gas Chromatography Column for the Analysis of Benzene and Toluene Gaseous Mixtures, Sensors and Actuators B: Chemical,Vol.141, No. 2, pp.431-435. ISSN 0925-4005.

R. L. Grob.(1977), Introduction,Modem Pracrice of Gas Chromatography. John Wiley & Sons, Inc., ISBN 0-471-59700-7, New York.

Radadia. A, R. I. Masel and M. A. Shannon.(2007). New Column Designs for Micro-GC,Transducers '07, Lyon, France, Jun. pp.10-14.

S.-I. Ohira, K. Toda, (2008).Micro gas analyzers for environmental and medical applications, Analytica Chimica Acta, Vol. 619,No. 2, pp. 143–156

Zampolli, I. Elmi, J. Stürmann, S. Nicoletti, L. Dori, G.C. Cardinali (2005). Selectivity enhancement of metal oxide gas sensors using a micromachined gas chromatographic column, Sensors and Actuators B: Chem, Vol. 105 ,No.2, pp. 400-406, ISSN 0925-4005.

3

Stationary Phases

Vasile Matei, Iulian Comănescu and Anca-Florentina Borcea
Petroleum and Gas University of Ploieşti, Romania

1. Introduction

Gas chromatography is a physical method of separating and identifying mixtures, with application in chemical practice, scientific investigation, petroleum technology, environmental pollution control, the food industry, pharmacology, biology and medicine. Separation of the sample components takes place by adsorption by a solid or by dissolving in a non-volatile liquid, called the stationary phase. Separation takes place based on different partitions of the sample components between a stationary phase and mobile phase. The mobile phase, in gaseous state, must be insoluble in the stationary phase (solid or liquid), the mobile phase continuously circulating over the stationary phase. Separation and measurement of components in the sample are made in a device called a gas chromatograph. The "heart" of the device is the chromatography column. The chromatography column contains the stationary phase, solid or liquid, as packing material for packed columns, or on the walls of capillary columns. This method of analysis has undergone important developments in recent decades and contributed to the progress of important scientific and applied fields.

2. Requirements of stationary phases

The principle of gas chromatography (GC) is based on the capacity of the stationary phase to produce different separation times upon exiting a chromatographic column that contains, under one form or another, stationary phases for the various mixture components of the sample. Through GC analysis, the sample components are separated through the combined effect of the stationary and mobile phases. The mobile phase is generally a gas, such as H_2, He, N_2, AR etc. The stationary phase is fixed and can produce either adsorption or absorption. A thermodynamic equilibrium is established between the two phases; this is expressed through the theoretic plates. When the sample components are added between the two phases at equilibrium, a difference emerges that can be expressed as a function of the quantities of sample components.

Chromatography is a method of separating multicomponent mixtures. This method relies on differences in partitioning behaviour between a flowing mobile phase and a stationary phase in order to separate the components in a mixture and on different velocities of compounds in the mobile phase. The mobile phase is usually a permanent gas, such as hydrogen, helium, nitrogen, argon etc., as a constant flow with a certain pressure. The stationary phase may be a solid or a liquid that is immobilized or adsorbed in a solid. The stationary phase may consist of particles (porous or solid), the walls of a tube (i.e. capillary) or a fibrous material.

Techniques by physical state of mobile and stationary phase:

Chromatography (C); Liquid chromatography (LC); Gas chromatography (GC); Paper chromatography (PC); Thin-layer chromatography (TLC) Column chromatography (CC) Gas liquid chromatography (GLC); Gas-solid chromatography (GSC); Liquid-solid chromatography (LSC); Size-exclusion chromatography (SEC); Liquid-liquid chromatography (LLC); Ion exchange chromatography (IEC).

Components separated from the sample are placed in a chromatographic column, a tube in stainless steel, copper, aluminium or glass for packed chromatography, or quartz for capillary chromatography. The column contains the stationary phase as a granular porous solid. The columns are in the oven of the chromatograph, isotherm or temperature programmed. The mobile phase passes through the column at a constant flow, the flow values being correlated with the column type (packed, capillary). Thus, a primary classification could be gas-solid chromatography and gas-liquid-solid chromatography.

Capillary columns are used mostly, but not exclusively. They are the latest and best method, but they cannot replace totally the packed columns. The trends in column chromatography are:

1. Packed columns are still used in 20% of chromatographic analysis.
2. Packed columns are primarily for preparative applications, permanent gas analysis and sample preparation.
3. Packed columns will be used in the future because some applications demand packed columns not capillary columns.

Gas chromatography has several advantages as a physical method of separating gas mixtures, namely:

1. Resolution. The technique is applicable to systems containing components with very close boiling points. By choosing a suitable stationary phase or adsorbent, molecules with similar physical and chemical properties could be separated. Sample components could form in normal distillation, azeotropic mixtures.
2. Sensitivity. The properties of gas chromatographic systems are responsible for their widespread use. The detector based on thermal conductivity of the component can detect picograms of sample. The sensitivity is important considering that a chromatography test is less than 30 minutes. Analysis that typically occurs in about one hour or more can be reduced to the order of minutes due to the high diffusion rate in the gas phase and the phase to fast equilibrium between mobile and stationary phases.
3. Convenience. Operation in gas chromatography is a direct operation. It does not require highly qualified personnel to perform routine separation.
4. Costs. Compared with many other now affordable analytical tools, gas chromatography presents excellent cost value.
5. Versatility. GC is adaptable, from samples containing permanent gas up to liquids having high boiling points and volatile solids.
6. High Separation Power. If you use mobile phases with a low viscosity degree, very long columns will provide a strong separability

7. Assortment of Sensitive Detecting Systems. Detectors used in gas chromatography are relatively simple, with high sensitivity and fast responses given.
8. Ease of Recording Data. Output of the gas chromatograph detector can be conveniently connected to a potentiometer recorder, or integrating systems, computers that can store a large amount of information.
9. Automatization. Gas chromatographs can be used to automatically monitor various chemical processes that allow samples to be taken periodically and injected into the chromatographic columns to be separated and detected.

An ideal stationary phase is selective and has different adsorptivity for each sample component in order to ensure separation, as well as a wide array of operating temperatures. It has to be chemically stable and have a low vapour pressure at high operating temperatures. Some criteria for selecting an adequate stationary phase are:

- Is the stationary phase selective enough in separating the sample components so as to separate them one by one?
- Can there be an irreversible chemical reaction and can the mixture components be separated?
- Does the liquid phase have a vapour pressure low enough at operating temperature?
- Is it thermally stable?

For instance, we quote a separability criterion of sample components known as "component sympathy" for the fixed phase.

Thus, a light n-paraffin mixture with "sympathy" for non-polar stationary phase will separate on grounds of different boiling points; olefins, being polarisable, will show "sympathy" towards a polar stationary phase. The cis- and trans- components of an olefin mixture will separate on a stationary phase consisting of a complex of transitional metals, dissolved in an appropriate solvent, such as polyethylene glycol.

Identifying mixture components can be accomplished using chromatographic etalons - pure substances or known mixtures of components. In these cases, universal detectors may be used (FID, WCD). For unknown mixtures, mass spectrometry is recommended. For the universal detector WCD, only non-fuel components may be used, and for the FID, only fuel components that do not interact chemically with the stationary phase may be used.

The quantitative analysis of a given component is based upon evaluating the chromatographic peak, which is triangle-shaped when columns with filling are used; its surface is measured and divided by the total surface, in different percentages for different types of detectors.

For capillary columns with good resolution, the signal takes the shape of straight lines and calculating the composition of the mixture is done in the order of succession, dividing each individual line by the total number of lines and using an adequate calibration curve, drawn upon determinations of known compounds.

The results can be expressed as a percentage, g/L component in a known mixture, depending on the type of calibration or the calibration units.

Thermal stability is of great importance, since operating under severe circumstances can lead to short analysis intervals and incomplete use of the chromatographic column; for instance, many stationary phases undergo degradation or decomposition at operating temperatures above 250^0 C, resulting in a lower operating life. On the other hand, certain components of unknown mixtures may reversibly or irreversibly poison the stationary phase components, deactivating it over time. In such cases, entry within the stationary phase can be accomplished using guarded columns that retain the poisonous components. It is vital that these components are known so as to prevent poisoning precious columns.

Before usage, the stationary phases placed into the GC columns must be conditioned and activated through streaming the mobile phase – the permanent gas.

Overall, stationary phases constitute a solvent with different selectivity for each component of the mixture in the analyzed sample.

2. Adsorption

Solid adsorbents as powdery material perform two functions in chromatography: the adsorbent itself (GS), in which case the separation takes place by adsorption, and support for stationary phase (non-volatile liquid), liquid stationary phase (GLS) coated on a granular material, in which case the separation takes place by absorption. Commonly, the size of the solid material used in chromatography is in the range of mesh (mesh - the number of openings per linear inch of a screen). Conversion between the Anglo Saxon and European expression (in mm) is presented in Table 1.

U.s. Mesh	Inches	Microns	Millimetres	U.s. Mesh	Inches	Microns	Millimetres
3	0.2650	6730	6.730	40	0.0165	400	0.400
4	0.1870	4760	4.760	45	0.0138	354	0.354
5	0.1570	4000	4.000	50	0.0117	297	0.297
6	0.1320	3360	3.360	60	0.0098	250	0.250
7	0.1110	2830	2.830	70	0.0083	210	0.210
8	0.0937	2380	2.380	80	0.0070	177	0.177
10	0.0787	2000	2.000	100	0.0059	149	0.149
12	0.0661	1680	1.680	120	0.0049	125	0.125
14	0.0555	1410	1.410	140	0.0041	105	0.105
16	0.0469	1190	1.190	170	0.0035	88	0.088
18	0.0394	1000	1.000	200	0.0029	74	0.074
20	0.0331	841	0.841	230	0.0024	63	0.063
25	0.0280	707	0.707	270	0.0021	53	0.053
30	0.0232	595	0.595	325	0.0017	44	0.044
35	0.0197	500	0.500	400	0.0015	37	0.037

Table 1. Mesh to millimetres conversion chart

Hereinafter we present the advantages of adsorption chromatography, which takes place at gas solid interaction, compared to GLS.

The advantages of gas chromatography gas-solid type from gas-liquid-solid can be summarized as follows: 1) stability of stationary phase on a wide range of temperatures; 2) a low detector noise limit; 3) lower values of HETP (height equivalent to a theoretical plate) than for gas-liquid chromatography (the adsorption-desorption process can be faster than the corresponding diffusion process in the liquid phase); 4) increased capacity of structural selectivity in separating geometrical isomers (e.g. using molecular sieves or graphitized carbon black); 5) high chemical selectivity when complexing agents are used as adsorbents; 6) high adsorption capacity allows the separation of gaseous or vapour compounds at room temperature; 7) increased chemical stability of adsorbents provides analysis of aggressive compounds; 8) important techniques have been developed in solids photochemistry and therefore there is better knowledge of adsorption; 9) numerous studies of heterogeneous catalysis were developed.

The limitations of this technique can be summarized as follows: 1) risk of asymmetric area due to the nonlinear adsorption isotherm of some analyzed compounds; 2) low reproducibility of chromatographic characteristics because adsorbents' properties are not as easily standardized as liquid substances; 3) more loss of the analyzed compounds as a result of irreversible adsorption or catalytic conversion of the separation process; 4) limited access to different commercial adsorbents for gas chromatography; 5) strong dependence of retention time of the sample size is common due to nonlinear adsorption, which combines the main advantage of gas-solid chromatography to the efficiency of capillary chromatography.

Adsorbents' selectivity is independent of the type of column used (packed or capillary column). Gas chromatography of adsorption has applications not only in the separation of gas mixtures with low boiling points, but also in the separation of hydrocarbon mixtures, as well as some organic compounds and some aromatic separation.

In the recent past there was a distinct tendency towards solid-gas chromatography. This can be attributed to two factors: first, new achievements in the field of adsorbents and second, using the improved logistics of gas chromatography (better trace analysis, more reproducible techniques for temperatures programming and high coupling techniques). Both solid and liquid stationary phases must have extraordinary properties, among which solid stationary phases are designed to solve many critical issues involved (Rotzsche, 1991). Adsorbents can be classified according to following criteria: chemical structure and geometric structure.

According to the chemical nature classification different interaction types of molecules' samples can take place. Kiselev & Yashin (1985) proposed grouping adsorbates into four groups (A-D), and in three groups (I-III) of adsorbents.

Adsorbents of type I have no functional groups or ions on the surface and thus they are not able to interact specifically with adsorbates. Interaction with all types of molecules A-D occurs non-specifically. Adsorbents are saturated hydrocarbons, graphite or rare gas crystals. The most important representatives of this type is graphitized carbon black (thermal graphitized carbon black) whose properties are close to ideal non-specific adsorbents. Similar to graphite are some inorganic compounds, such as graphite-like boron nitride (BN) or sulphides of metals (i.e. MoS) (Avgul et al., 1975).

Adsorbents of type II develop positive partial charge on the surface adsorbent. Besides these dispersion forces specific interactions develop leading to the orientation and localization of adsorbate molecules on centres having high charges. That concerns the salts having cations with positive charge with small ray, while negative charges are distributed in a relatively large volume (i.e. $BaSO_4$). Thus, the most significant representatives of this type are adsorbents with functional groups of protonated acids, such as hydroxylated silicagel, Lewis acids aprotic centres located on the surface. Molecules A type (saturated, rare gases) are non-specifically adsorbed and they are dispersion forces only. Molecules B, C and D can be adsorbed non-specifically. Type B includes molecules with electron density localized on some bonds or atoms: type II bonds of unsaturated or aromatic compounds, functional groups, atoms having electrons couple (ethers, ketones, tertiary amines, pyridines and nitriles), molecules with high quadrupole moments (N_2 molecules). Interactions between an adsorbent type II and an adsorbed type B develop between centres with electronic high-density (molecules in the sample) and positive charges of adsorbent (i.e. acidic proton of hydroxylated silica gel or a cation of Li, Na, Mg, Ca in acid zeolite or aprotic acids Lewis type on the surface). Type C molecules have a positive charge located on a metal atom and the excess of electron density is distributed on adjacent bonds (organometallic compounds). Type D are molecules containing peripheral functional groups (OH, NH, etc.) whose electronic density increase the density of some atoms (O, N) and reduce the density for other atoms (H). This group includes water, alcohols and primary and secondary amines. Specific interactions of type D adsorbates with Type II adsorbents involves the forces between the adsorbent centres with positive charges and the only couple of electrons of the atoms of O or N of the sample molecules.

Adsorbents type III. Adsorbents types III are specific having centres with high density centres on the surface. In this group are polymers as polyacrylonitrile, copolymers of vinylpyridine and divinylbenzene and polymers group with (C = O) and (–O–) on the surface. This group could include porous polymers based on styrene ethylvinylbenzene cross linked with divinylbenzene, varying by using different polymerization promoters and even non-specific dispersion forces. Adsorbents Type III includes crystalline surfaces formed by anions and chemically modified non-specific adsorbents covered with a monomolecular layer of adequate substance, creating negatively charged surface centres. Adsorbents type III interacts non-specific with adsorbates type A and specific with type B, C or D by forces from the negative charges of the adsorbent surface and from the positive charges of metallic atoms (C) or from functional groups (OH, NH) proton type (D) and dipole or induced dipole type (B).

Classification according to geometric structure

This classification concerns the possibilities to increase the surface. Increasing the surface meets a series of reserves, such as an increase in surface leads to increased dispersion with an increase in heterogeneity. Therefore, increasing contact points between particles will reduce the pore diameter with Knudsen diffusion disadvantage. Kiselev & Yashin (1985) have overcome these difficulties in the development of GCS. The adsorbents are geometrically classified as follows (Rotzsche, 1991).

Type 1 Non-porous adsorbents

Crystalline products with a smooth surface (sodium chloride, graphitized thermal carbon black, BN, MoS,). S_A in the range of 0.1 – 12 m^2/g.

Type 2 Uniformly porous adsorbents with wide pores. Silica gel with pore diameters between 10 and 200 nm (Porasil, Spherasil) and some other forms of silica gel chemical bonded (Durapak a.o.) and other polymers styrene divinylbenzene type with large pores (20 - 400 nm).

Type 3 includes adsorbents with a uniform pores system, but small pores. In this category should be mentioned molecular sieves (zeolites), carbon molecular sieves, porous glass and porous polymers. Pores diameter is around 10 nm.

Type 4 presents adsorbents with non-uniform pores. Among materials of this type there are activated carbon and alumina. Because of geometrical and chemical heterogeneity, having pores in the range 2-20 nm (mezopores) up to 200 nm, these adsorbents are not suitable for GCS. Classification is based on pores size. The difference between porous adsorbents and non-porous adsorbents consists in the free spaces form developed by the porous system. This system is quantitatively characterized by the following parameters (Rotzsche 1991):

- Specific surface area S_A (geometric size of the pore wall area/gram of adsorbent).
- Specific pore volume V, (total pore volume/gram of adsorbent).
- Mean pore diameter d_{50} (average diameter of 50% of the pores).
- Pore size distribution.

These parameters could be measured by gas chromatography and other methods, such are mercury porosimetry and reversed size exclusion chromatography. The ratio of pore diameters for porous adsorbents to the molecules diameters of the adsorbate is a significant parameter because the higher ratio is (the molecule diameters are smaller compared with pore diameters), the faster the adsorption equilibrium is reached. For a similar size of pore diameters and molecules diameters the adsorption rate depends equally on the pore shape and the adsorbate molecules' size. For narrow pores the adsorbed molecule on surface atoms of one pore could interact with other pore surface atoms and in this case the adsorbed molecules are trapped, and the transfer of molecules between adsorbent and mobile phase is stopped.

The most widely used chromatography supports are Kieselguhr (diatomaceous earth = light-coloured porous rock composed of the shells of diatoms) based, containing polysilicic acid in the hydrated amorphous silica form having a porous structure and containing varying amounts of metal oxides of Fe, Al, Mg, Ca, Na, K. These supports are known under different commercial terms, the most commonly used type being *Chromosorb*. They are found in several varieties, of which the most important are:

- *Chromosorb P* (pink) is reddish pink and has a surface area situated between 4 -6 m^2 / g. It has the advantage that it can be loaded with a large amount of liquid stationary phase (up to 25 30%) and has high mechanical strength. It is not sufficiently chemically inert and for this reason it is used mainly to separate non-polar substances.
- *Chromosorb W* (white) is white and has a smaller surface area than Chromosorb P, 1-2 m^2 / g. Therefore it can be loaded with a smaller amount of liquid stationary phase, up to 15%. It is chemically inert, but is friable, having a low mechanical strength. It has general usability, so it can be used to separate all classes of compounds, both polar and non-polar.
- *Chromosorb G* surface area is even smaller than Chromosorb W, of only 0.5 m^2 / g and can upload up to 5% stationary phase. But it has the advantage of combining the chemical inaction of type W with the mechanical strength of type P.

There are also supports of other types, but these are much less used. Such supports are based on synthetic polymers (i.e. Teflon), glass and others.

Next will be presented several examples of a few illustrative applications of some adsorbents (Poole, 2003) Table 2:

Stationary Phase	Maximum Temperature (°C)	Usual Applications
Alumina	200	Alkanes, alkenes, alkines, aromatic hydrocarbons - from C_1 to C_{10}
Silica gel	250	Hydrocarbons (C_1 - C_4), inorganic gases, volatile ethers, esther and ketone
Carbon	350	Inorganic gases, hydrocarbons (C_1-C_5)
Carbon molecular sieves	150	Oxygenated compounds (C_1- C_6)
Molecular sieves (5X and 13X)	350	Hydrogen, oxygen, nitrogen, methane, noble gases. Separation He/Ar and Ar/O_2. Hydrocarbons C_1 - C_3 on 3X, on 13X till C_{12} but no isomers. Cyclodextrine, halocarbons, permanent gases, hydrofluorocarbons, hydrocarbons C_1-C_{10}.
Q	310	Hydrocarbons C_1-C_{10}, halocarbons C_1- C_2.
S	250	Volatile organic solvents C_1 - C_6
U	190	Nitrocompounds, nitrils, water, inorganic gases.

Table 2. Illustrative examples of some adsorbents and temperature values at which they are active

Table 3 briefly presents some gas chromatography main adsorbents' characteristics (Grob & Barry, 2004).

Adsorbent	Polymeric material	Tmax, °C	Applications
HayeSep A	DVB-EGDMA	165	Permanent gases, including: hydrogen, nitrogen, oxygen, argon, CO, and NO at ambient temperature; can separate C2 hydrocarbons, hydrogen sulphide and water at elevated temperatures
HayeSep B	DVB-PEI	190	C1 and C2 amines; trace amounts of ammonia and water

Adsorbent	Polymeric material	Tmax, °C	Applications
HayeSep C	ACN-DVB	250	Analysis of polar gases (HCN, ammonia, hydrogen sulphide) and water
HayeSep D	High purity DVB	290	Separation of CO and carbon dioxide from room air at ambient temperature; elutes acetylene before other C2 hydrocarbons; analyses of water and hydrogen sulphide
Porapak N	DVB-EVB-	190	Separation of ammonia, carbon EGDMA dioxide, water, and separation of 165 acetylene from other C2 hydrocarbons
Porapak P	Styrene-DVB	250	Separation of a wide variety of alcohols, glycols, and carbonyl analytes
Porapak Q	EVB-DVB copolymer	250	Most widely used; separation of hydrocarbons, organic analytes in water and oxides of nitrogen
Porapak R	Vinyl pyrollidone (PM)	250	Separation of normal and branched alcohols
Porapak T	EGDMA (PM)	190	Highest-polarity Porapak; offers greatest water retention; determination of formaldehyde in water
Chromosorb 101	Styrene-DVB	275	Separation of fatty acids, alcohols, glycols, esters, ketones, aldehydes, ethers and hydrocarbons.
Chromosorb 102	Styrene-DVB	250	Separation of volatile organics and permanent gases; no peak tailing for water and alcohols
Chromosorb 103	Cross-linked PS	275	Separation of basic compounds, such as amines and ammonia; useful for separation of amides, hydrazines, alcohols, aldehydes and ketones
Chromosorb 104	ACN-DVB	250	Nitriles, nitroparaffins, hydrogen sulphide, ammonia, sulphur dioxide, carbon dioxide, chloride, vinyl chloride, trace water content in solvents
Chromosorb 105	Crosslinked polyaromatic	250	Separation of aqueous solutions of formaldehyde, separation of acetylene from lower hydrocarbons and various classes of organics with boiling points up to 200° C

Key: DVB – divinylbenzene; EGDMA – ethylene glycol dimethacrylate; PEI – polyethyleneimine; ACN – acrylonitrile; EVB – ethylvinylbenzene.

Table 3. Polymer type adsorbents

3. Types of stationary phase for adsorption

The conditions that liquid stationary phases have to meet are:

- To be a good solvent for the components of the sample, but the solubility of these components has to be differentiated;
- To be practically non-volatile at the temperature of the column (vapour pressure to be less than 0.1 mm Hg);
- Be chemically inert;
- Have a higher thermal stability.

The categories of the stationary liquid phase are:

- *Non-polar stationary phases*, which are compounds, such as hydrocarbons (paraffin) or silicone oils (polysiloxanes) without grafted polar groups. Examples of such stationary phases are: squalane ($C_{30}H_{62}$ hydrocarbon), silicone oils metylsilicon type (names OV 1, SE-30). These phases separate the analyzed compounds in order of increasing boiling points.
- *Polar stationary phases*, which contain a high proportion of polar groups, i.e. the average molecular weight polyethylene glycol (Carbowax 20M), silicone oils with cianopropil groups (OV 225 cianopropyl-methyl-phenyl silicone) diethylene glycol succinate (DEGS), nitrotereftalic ester of polyethylene glycol (FFAP) etc. They differentiate between the non-polar and polar compounds, retaining only those which are polar, and are used especially to separate polar compounds.
- *Intermediate polarity stationary phase*, containing polar groups in the lower concentration or polarizable groups grafted onto a non-polar support. Examples of such stationary phases are phenyl methyl silicone phase (OV 17), dinonyl phthalate, polyethylene glycols having high molecular weight. They are universal stationary phases, which can be used to analyze both polar and non-polar compounds.
- *Specific stationary phases*, which are used in certain cases. They contain compounds that interact only with certain components of the mixture to be analyzed, for example $AgNO_3$ dissolved in polyethylene glycol which forms adduct with olefins.
- *Chiral stationary phases*, containing chiral compounds interacting with only one optical isomer of a pair of enantiomers. Such phases are based on cyclodextrin or certain amino acids.
- Given the polarity of the stationary phase and possible interactions, organic compounds can be grouped in terms of chromatographic separation in the following five classes:
- *Class I* very polar compounds, able to give hydrogen bonds: water, glycerin, glycols, hydroxyacids, and amino acids. These compounds are difficult to separate by gas chromatography, due to high polarity. Excepting water, they derivatized before separation.
- *Class II* polar compounds, which have active hydrogen atoms: alcohols, carboxylic acids, phenols, primary and secondary amines, nitro and nitriles with a hydrogen atom in α position. These compounds are separated in polar stationary phases.
- *Class III* intermediate polarity compounds, without active hydrogen: ethers, esters, aldehydes, ketones, nitro and nitriles without hydrogen atom in α position. These compounds are separated on stationary phases of intermediate polarity.
- *Class IV* compounds with low polarity, but which have active hydrogen: aromatic hydrocarbons, alkenes, chloroform, methylene chloride, dichloroethane, trichloroethane

etc. These compounds are separated on stationary phases of intermediate polarity or non-polar.

- *Class V* non-polar compounds: alkanes, cycloalkanes. These compounds separate in non-polar stationary phases.

This classification is empirical and is only meant to facilitate choosing the most suitable stationary phase for chromatographic analysis. Many polymers contain mixtures of the above functional groups, as indicated in Table 4.

Name	Type	Structure	Density g/ml	Viscosity cP	Average Molecular Weight
OV-1	Dimethysiloxane gum	CH_3	0,975		> 10^4
OV-101	Dimethylsiloxane fluid	CH_3		1500	30000
OV-7	Phenylmethyl dimethyl Siloxane	80 %CH_3 20%C_6H_5	1.021	500	30000
OV-210	Trifluoropropyl methylsiloxane	50%$CH_2CH_2CF_3$ 50%CH_3	1.284	10.000	200000
OV-275	Dicyanoalkyl siloxane			20.000	5000

Table 4. Properties of some commercially available pol-siloxane phases

4. Kovats retention index

A universal approach (Kovats) solved the problems concerning the use, comparison and characterization of gas chromatography retention data. This reporting of retention data as retention time t_R is absolutely meaningless, because all chromatographic parameters and any experimental fluctuation affect the proper measurement of a retention time. Using relative retention data ($\alpha = t'_{R2}/t'_{R1}$) provided improvements, but the default of a universal standard usable for a wide range of temperatures for stationary phases of different polarity has discouraged its use. In the Kovats approach, to retention index I of an alkane is assigned a value equal to 100 times its number of carbon atoms. Therefore, for example, values I of n-octane, n-decane and n-dodecanese, are equal to 800, 1000, 1200, respectively, by definition, and are applicable to any columns, packed or capillary and to any mobile phase, independent of chromatographic condition, including column temperature. However, for any component, the chromatographic conditions, such as stationary phase, its concentration, support and column temperature for packed columns, must be specified. Since retention index is the preferred method of reporting retention data for capillary columns, stationary phase, film thickness and column temperature must also be specified for any component out of n-alkanes, otherwise I values are meaningless. A value I of a component can be determined by mixing a mixture of alkanes with the desired component and chromatography under specified conditions. A plot of the logarithmic adjusted retention time versus retention index is generated and a retention index of the solute considered is determined by extrapolation, as shown in the figure below for isoamyl acetate.

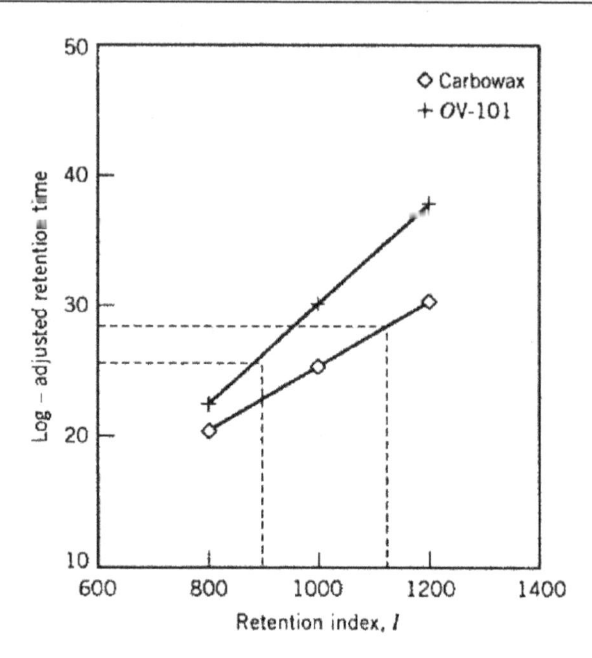

Fig. 1. Plot of logarithm-adjusted retention time versus Kovats retention index: isoamyl acetate at 120◦C (Grob & Barry, 2004).

Selectivity of stationary phases can be determined by comparing the values I of a solute on a non-polar phase, such as squalane OV 101 (I = 872) with corresponding I values of 1128 associated with more polar columns containing Carbowax 20M, for example. This difference of 256 units shows the superior retention of the Carbowax 20M column. Specifically, isoamyl acetate elutes between n-C11 and n-C12 on a Carbowax 20 M columns, but much faster on OV 101, where it elutes after n-octane. Retention indices in GC normalize instrumental variables, allowing retention data obtained by various systems to be compared. For example, isoamyl acetate with a retention index of 1128 will elute between C11 and C23 in the same chromatographic conditions. Retention indices are also very useful for comparing the relative elution order of the series of analytes on a specific column to a given temperature, and comparing the selective behaviour of two or more columns.

5. McReynolds classification of stationary phases

As shown, the most important criteria upon which stationary phases choice is made for a certain analysis is the polarity of the phase and of analyzed compounds. Numerical expression of this polarity, which would facilitate the choice, is not possible because there is no physical size that can be associated directly with the polarity (to some extent the dipole moment could be this size). For this reason, to express the stationary phase polarity they are using some reference compounds, which are arbitrarily chosen. Such a system was developed by Rohrschneider and later developed by McReynolds, based on the Kovats retention indices and using squalane as reference stationary phase, assigning it null polarity. A total of five reference compounds were selected: benzene, 1-butanol, 2-pentanona, nitropropane and pyridine. Each of these compounds can be considered standard for certain classes of substances, which are similar in terms of chromatographic behaviour, as follows:

- *benzene* for unsaturated hydrocarbons and aromatic hydrocarbons;
- *1-butanol* for alcohols, phenols, carboxylic acids;
- *2-pentanona* for aldehydes, ketones, ethers, esters;
- *nitropropane* for nitroderivates and nitriles;
- *pyridine* for aromatic bases and heterocycles with nitrogen.

The steps to determinate the constants McReynolds for a particular stationary phase A, for packed columns, are as follows:

- each reference compound is analyzed on a column with 20% squalane as stationary phase, isothermal at 100 ° C and the corresponding retention indices are determinate;
- reference compounds are analyzed on a column containing 20% stationary phase A, under the same conditions and in such cases Kovats retention indices are established;
- McReynolds constants of the stationary phase A are calculated as follows:
- for benzene (denoted by x '): the difference between the retention index of benzene on stationary phase A and on squalane:

$$x' = I_R(phase\ A) - I_R(squalane) \quad (1)$$

- for a butanol (denoted by y '): the difference between the retention index on phase butanol on stationary phase A and on squalane;
- The same applies for 2-pentanona (z'), nitropropane (u') and pyridine (s').

These McReynolds constants values were determined for a great number of common stationary phases and they are tabulated. Some values are also given in Table 5.

Name of the stationary phase	X'	Y'	Z'	U'	S'
Squalane $C_{30}H_{62}$	0	0	0	0	0
Methylsilicone OV-1	16	55	44	64	42
Methylsilicone SE-30	15	53	44	64	41
Methyl-phenyl-silicone (20% phenyl) OV-7	69	113	111	171	128
Methyl-phenyl-silicone (50%phenyl) OV-17	119	158	162	243	202
Cyanopropyl-methyl-phenyl-silicone OV-225	228	369	338	492	386
Carbowax 20M (Polyethylenglycol)	322	536	368	572	510
Nitroterephtalic ester of PEG (FFAP)	340	580	397	602	627
Diethyleneglycol succinate (DEGS)	496	746	590	837	835

Table 5. McReynolds constants values for some usual stationary phases

Currently there are hundreds of stationary phases used in gas chromatography and McReynolds constants allow selecting the one that promises the best separation of the components analyzed. Because obviously there are a number of stationary phases that have similar McReynolds constant values, they can be substituted in between without affecting separation. Thus, if there is a stationary phase recommended in the literature for separation, but is unavailable in a laboratory, it will be replaced with an equivalent.

There was an impulse of consolidating the number of stationary phases used in the mid-1970s. Leary et al. (1973) reported the application of the statistical technique *the nearest neighbour* for the 226 of the stationary phases of McReynolds study and suggested that a total of only 12 phases could replace the 226. Later it has been found that four phases, OV-101, OV-17, OV-225 and Carbowax 20M could provide GC analysis, satisfying 80% of a wide variety of organic compounds, and a list of six favourite stationary phases on which almost all gas-liquid chromatographic analysis could be performed: (1) dimethylpolysiloxane (OV-101, SE-30, SP-210); (2) 50% phenyl-polysiloxane (OV-17, SP-2250); (3) poly-ethylene-glycol of molecular weight> 4000 (Carbowax); (4) diethyleneglycol succinate (DEGS); (5) 3- ciano-propyl-polysiloxane (Silar-10 C, SP-2340) and (6) tri-fluor-propyl-polysiloxane (OV-210, SP-2401).

Another quality of McReynolds constants is guiding the selection of columns to separate compounds with different functional groups, such as ketones by alcohols, ethers by olefins and esters by nitriles. If the analyzer wants a column to elute an ester after an alcohol, the stationary phase should have a value Z' greater than the value Y'. Also, a stationary phase should have a value of Y′ greater than Z' in order to make ether elute before the alcohol.

6. Column type

Different chromatography types of column are presented by their diameters. Thus, there are three main types of column: packed columns, capillary columns and micro-packed columns.

Packed columns

The most used column materials are stainless steel, glass, quartz, nickel and polytetrafluorethylene (PTFE). Because of their good thermal conductivity and easy manipulation, stainless steel columns are widely applied. Glass columns are more inert. PTFE columns are used to separate mixtures of halogens and derivates. Classical packed columns′ inner diameters range from **1.5** mm to **6** mm. Very important for packed columns is the load; the load is directly proportional to the column cross-section and hence to the square of the inner diameter (Leibnitz & Struppe, 1984).

Capillary columns (open tubular columns)

The terminology used for capillary column includes names such as WCOT (well coated open tubular) which are capillary columns where the stationary phase is deposited directly onto the inner surface of the wall not including substances that could be considered support. SCOT (support coated open tubular) includes those capillaries that have deposited on the inner surface a finely divided support (NaCl, BaCO3, SiO2, etc.) in which stationary phase is submitted. PLOT (porous layer open tubular) columns contain capillaries with stationary phase deposited on a porous support consisting of a finely divided support for packed gas chromatography columns (Celite, Chromosorb, etc.). Porous material is deposited in the capillary tubes from the very first moment of column conditioning. Capillary columns BP, DP, SPB (bond phase) include capillary tubes that have stationary phase chemically immobilized on the surface of the capillary inner wall. When choosing a certain type of capillary column it is necessary to take into account the nature of the sample being analyzed, the equipment available and the effective separation. Capillary columns WCOT have better efficiency in separation than SCOT because the inner column diameter is smaller. SCOT columns can be considered as subgroups of PLOT columns, since all SCOT are PLOT columns, but not vice versa (Ciucanu, 1990). For capillary columns the inner diameter ranges from 0.10 to 0.20 mm.

Wall-coated open-tubular (WCOT) columns contain the stationary phase as a film deposited on the internal surface of the tube wall. The formation of a thin and uniform film along the total length of the column represents the key to capillary columns' efficiency.

Porous-Layer Open Tubular (PLOT)

To increase the sample capacity of capillary columns and decrease the film thickness, porous layers for the inside walls of the column tubing were used. The stationary phase amount is directly related to the efficiency of capillary columns and generates efficiency increasing due to porous layers.

The method presented here used capillary columns of glass and quartz, the most used today. Glass capillaries are drawn from a device consisting of an electric furnace in which glass melts, two traction rollers which pull the column, a curved metallic tube for the spiral of the tube electrically heated at glass melting temperature. Capillary dimensions are set from the ratio of the two rolls' traction speed. Silica (quartz) tubes require temperatures of 1800-2000 ° C. After pulling, the capillary tube is covered with a polymer film in order to increase the tensile strength. Polyamide or polyimide polymers are used, which are stable up to temperatures of 350-400 ° C.

For the glass preparation, silica mixed with some metal oxides (Na_2O, CaO, MgO, B_2O_3, Al_2O_3 etc.) are used that break covalent bonds. In general, glass is chemically inert; however, because of impurities and the superficial structure, capillaries present catalytic activity and adsorption of the sample component. Metal oxides will act as Lewis acid. Molecules of analyzed compounds containing π electrons or non-participating electrons will react with Lewis acids. Adsorption and catalytic properties of glass and silica are due to silanol groups and siloxane bridges on the surface of the capillary tube. Also, water in the atmosphere adsorbs to the surface by hydrogen bonds. Heating the tube, water can be removed in order to form stable siloxane bridges. For a capillary tube to become a chromatographic column, the inner surface of the tube must be covered with a uniform and homogenous film of stationary phase.

By silanization of silanol groups, the critical surface tension (one that is established between the solid surface and stationary phase) is significantly reduced, but has the same value for glass and quartz. Silanizated surfaces have a surface tension so small that only non-polar stationary phases (OV-101) can "wet" them. Polar stationary phases (Carbowax 20M) do not give uniform films. The basic elements of glass capillary tubes' chemistry are alkaline metal ions, silanol groups and siloxane bridges. These are active centres that catalyze at high temperatures the decomposition reactions of sample components as well as of the stationary phase. For these reasons these active centres have to be deactivated.

Capillary tube surface deactivation involves removing the metal ions and blocking out the silanol groups with no catalytic activity groups. Removal of metal ions of silica and glass capillary tubes is achieved by acid washing in a static or dynamic regime. An acidic wash causes not only metal removing, but also increases the number of silanol groups, by breaking the siloxane bonds on the surface of the tube. In the past, tubes' surface deactivation was with surfactants, but they have low thermal stability and are no longer used. Current methods use disilazane, cyclic siloxanes and polysiloxanes. In all cases siloxane groups turn into silicate esters. Methods to deactivate with silazane realize the blocking of silanol groups by reaction with hexamethyldisilazane, di-fenil-mehyl-disalazane, dibutiltetramethyldisilazane, dihexyltetramethyldisilazane etc.

Reaction for deactivation with disilazane is:

$$Si - OH + NH(SiR_3)_2 \rightarrow Si - O - SiR_3 + NH_2 - SiR_3$$

$$Si - OH + NH_2 -SiR_3 \rightarrow Si - O - SiR_3 + NH_3$$

Uniform films of stationary phase can be obtained when the adhesion forces between stationary phase and capillary column surface are greater than the stationary phase molecules. Cohesion forces are stronger with increased similarity between organosiloxanic groups and functional groups of stationary phase. Methyl groups have high thermal stability, are chemically inert and are used to deactivate capillary columns with non-polar stationary phase. Phenyl groups have lower thermal stability than methyl groups, but allow the deposition of medium polarity stationary phase. Cyanopropyl and trifluoropropyl groups are strong polar and it is difficult to bond on the glass by reaction with disilazane, because disilazane have a reduced thermal stability. Another method of blocking the activity of silanol groups is based on reaction with cyclosiloxanes. Deactivation with cyclosiloxanes allows bonding some functional groups similar to those of the stationary phase.

Another method for modifying the surface chemistry of glass and silica capillary tubes is based on the stationary phase, by blocking the surface active centres, as well as getting a very stable film of stationary phase. Carbowax 20M was the first stationary phase used to deactivate glass capillary tubes. This step is followed by a heating treatment.

Another method of decomposition of the stationary phase is to use radicalic initiators. The initiator used for immobilization of polyethylene glycol is dicumilperoxide. Another possibility of immobilization of Carbowax 20 M is based on the reaction between OH groups of polyethylene glycol with diisocyanate in the presence of dibutyltindilauril. A chemically inert and thermal stability capillary column improves also by immobilizing the stationary phase polysiloxanes type with terminal hydroxyl groups OH, such as OV-17-OH, OV-31-OH, OV-240-OH.

Changing the surface chemistry by immobilizing the stationary phase can be done with a large number of siloxane polymers. A simple method of immobilization is polymerization in a column of silicon monomers consisting in the formation of bonds Si–O–Si between siloxane polymers and the capillary column wall. Immobilization can be done in the presence of radicalic initiators. Immobilization of the stationary phase can be initiated also by gamma radiation at room temperature. Stationary phases without vinyl groups require a high dose of radiation. Immobilization in the presence of gamma radiation eliminates the danger of degradation of the stationary phase and allows very precise control of reaction.

Micro-Packed Columns

Due to their parameters concerning mass capacity, phase ratio, resolution and short analysis time, micro-packed columns are very attractive. They are micro pore tubes having inner diameters ranging from 0.3 - 1 mm and at lengths varying from 1 to 15 m, packed with particles 0.007-0.3 mm in diameter. Table 6 presents geometrical characteristics for packed columns and micro-packed columns.

	Irregular micro-packed column	Classical packed columns	Regular micro-packed columns
Tube i.d.(*d,*) [mm]	0.3-0.5	≥ 2	0.3-1.5
Particle size (*d,*) mm	0.05-0.15	0.12-0.30	0.04-0.3
d_p/d_c	0.04-0.10	0.07-0.25	0.25-0.5
sample capacity µg	5-20	>1000	1-10
h (mm)	0.4-1.2	>0.5	0.15-0.40
L ,m	1.8 (6 ft)	4.5 (15 ft)	4.5 (15 ft)
Preferred applications	Short columns permit high-speed analyses coupled techniques with MS.	Simple separation problems; trace analysis; coupled techniques with spectrometry	Multi-component analysis; high-resolution coupled technique with MS; trace analysis

Table 6. Geometrical characteristics for packed columns and micro-packed columns

Micro-packed columns are suitable for the analysis of multi-component mixtures and trace constituents, especially when coupling with mass spectrometry, thus allowing quickly analyses.

7. Operation column

Several parameters can be used to evaluate the operation of a column and to obtain information about a specific system. An ideal gas chromatographic column is considered to have high resolving power, high speed of operation and high capacity.

Column Efficiency

Two methods are available for expressing the efficiency of a column in terms of HETP (height equivalent to a theoretical plate): measurement of the peak width (1) at the baseline, $N=16(V_R/w_b)^2$ and the peak width at half the peak height (2), $N=5.54(VR/w_k)^2$, where N is the height equivalent to theoretical plate (HETP) and w has the significance from the graphic below (Figure 2).

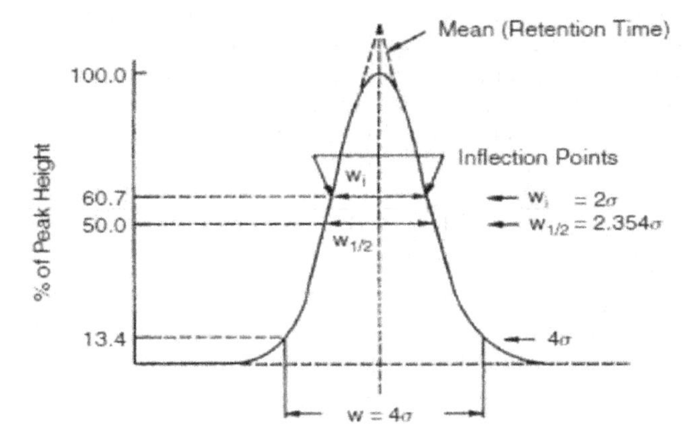

Fig. 2. Methods for expressing the efficiency of a column in terms of HETP

Effective Number of Theoretical Plates

Open-tubular columns generally have a larger number of theoretical plates. The *effective number of theoretical plates*, N_{eff}, characterizes open-tubular columns.

$$N_{eff}=16(V_R'/w_b)^2 \quad (2)$$

where V'_R the adjusted retention volume.

Separation Factor

The *separation factor* (S) describes the efficiency of open-tubular columns:

$$S= 16(V_R'/w_b)^2 =16(t'_R /w_b)^2 \quad (3)$$

where V'_R is the adjusted retention volume and t'_R is the adjusted retention time.

8. Preparation of columns

The most used technique for coating supports with high concentrations (>15%) of viscous phases is solvent evaporation. This technique leads to a uniform phase deposition. The steps to follow are presented below (Grob & Barry, 2004).

1. Prepare a solution of known concentration of liquid phase in the suitable solvent.
2. Add the desired amount of solid support in a known volume of the solution.
3. Transfer the mixture in a Büchner funnel, in order to remove the solvent excess.
4. Measure the volume of filtrate.
5. Dry the "wet" packing to remove residual solvent.
6. Calculate the mass of liquid phase retained on the support.

This technique allows obtaining a uniform coating of a support and minimizes the oxidation of the stationary phase during column.

9. The technology of capillary columns

The widespread use of fused silica for capillary columns is because it is inert compared to other glasses. In 1986 Jennings published a comparison of fused silica with other glasses, such as soda-lime, borosilicates and lead. Fused silica is formed by introducing pure silicon tetrachloride into a high-temperature flame followed by reaction with the water vapour generated in the combustion (Jennings, 1997). On the surface they have distinguished three types of silanol groups: adsorptive, strong, weak or none. The composition of some glass and silica are presented in Table 7.

Glass	SiO_2	Al_2O_3	Na_2O	K_2O	CaO	MgO	B_2O_3	PbO	BaO
Soda-lime	68	3	15	-	6	4	2	-	2
Borosilicate (Pyrex 7740)	81	2	4	-	-	-	13	-	-
Potash soda-lead (Corning 120)	56	2	4	9	-	-	-	29	-
Fused silica	100								

Table 7. Composition of some glass and silica (Jennings, 1986)

10. Column performances

Column performances need to take into account van Deemter expression, carrier gas choice, effect of carrier gas viscosity on linear velocity and phase ratio. Column efficiency, the resolution and sample capacity of a capillary column depend on construction material of the column, the inner diameter and the film thickness of the stationary phase. The distribution coefficient K_D could be interpreted as a function of chromatographic parameters. For a given temperature column of a tandem solute-stationary phase, K_D is constant. In this situation, K_D represents the ratio between the concentration of solute in stationary phase and the concentration of solute in carrier gas:

$$K_D = k\beta = r/2d_f \quad (3)$$

where $\beta = r/2d_f$ is the phase ratio, r is the radius of the column and d_f is the film thickness of stationary phase.

11. Selecting stationary phases

Using packed columns for gas chromatography needs to have a "collection" of available columns packed with different stationary phases and variable lengths for different purposes. The non-volatile liquid phases for packed columns have a good selectivity, but the thermal stability as well as the difference of thickness of the film at elevated temperatures is low and this will affect the inertness of stationary phase and the efficiency of the column.

For the best choice of a stationary phase it is important to remember the motto "like dissolves like". A polar phase could lead to a lower column efficiency compared to a non-polar phase. The maximum temperature limits decrease, as well as the operation lifetime if operating temperature is high (Grob & Barry, 2004). The lower thermal stability of polar phases can be avoided using a thinner film of stationary phase and a shorter column for lower elution temperatures.

12. Installing, conditioning, exhausting and regeneration of chromatographic columns

On column installing it is important to select the proper ferrule, meaning that the inner diameter of the ferrule fits to the outer diameter of the column, in order to avoid loss of carrier gas. Table 8 presents the size of ferrules suitable to column size (inner diameter).

Ferrules inner diameter, mm	Columns inner diameter, mm
0.4	0.25
0.5	0.32
0.8	0.53

Table 8. Ferrules inner diameter compatible with columns inner diameter

In gas chromatograph the chromatography column is placed in the oven, after the injector and before the detector. Between the injector and the column a *retention gap* or *guard column* is often installed. Its length is 0.5-5.0 m and it is actually a deactivated fused-silica tube. In this retention gap the condensed solvent resides after injection, but it is removed by vaporization. The role of a guard column is to collect the non-volatile compounds and particulate matter from samples in order to avoid their penetration in the chromatography column and maintain the lifetime of the column.

Conditioning of chromatography columns is recommended for residual volatiles' removal. For capillary column conditioning there are three steps to follow:

1. Carrier gas flow has to be constant when column temperature is higher than room temperature.
2. The maximum temperature of the stationary phase or of the column must not be transcended because the chromatography column could be damaged.
3. For capillary column, in order to obtain a steady baseline, it is indicated only to purge carrier gas flow for 30 min at ambient temperature and then, with a rate around 4°C/min, elevate the temperature slightly over the maximum temperature limit and maintain for several hours. The first two procedures are also applicable for conditioning of a packed column.

When non-volatiles and particulate matter are accumulating on the inlet of the column or on the injector liner, the chromatography column is contaminated and it is no longer working to its full potential. This is reflected mostly in peak tailing and changes in the retention characteristics of the column. Column reactivation could be realized either by removing the inlet part of the column (1-2 meters) or by turning the column around and then applying step 3 for a longer time. The extreme solution is to remove the solvent.

13. Practical application of GC analysis for mixtures of hydrocarbons

One of the fields where GC is increasingly used is in the oil and gas industry, and specially the refinery sector. In refineries the GC is used not only for identifying the components from a mixture, and then concentration, but also as an integrate tool for loop control.

The gas mixture that you may find on a refinery is usually called "refinery gas", and is a mixture of various gas streams produced in refinery processes. It can be used as a fuel gas, a final product, or a feedstock for further processing. An exact and fast analysis of the components is essential for optimizing refinery processes and controlling product quality. Refinery gas stream composition is very complex, typically containing hydrocarbons, permanent gases and sulphur.

In this practical example of GC use in a refinery, the hydrocarbon mixture was an industrial C5 cut separated from a Fluidized Catalytic Cracking (FCC) plant. Table 9 (Comanescu & Filotti, 2010).

The gas chromatograph was a HP 5890 Series II apparatus, with a configuration and operating parameters similar to those for PONA analysis (ASTM D 6293-98). Commercially available standard (Agilent) for refinery gas analysis was employed for calibration of both hydrocarbons retention times and concentrations.

Component	Normal boiling point, °C	Raw refinery C5 cut (approx. average) % mass
C4 hydrocarbons	- - -	1.88
isopentan	27.85	45.76
1-pentene (α-amylene)	29.97	5.37
n-pentane	36.07	8.36
isoprene	34.07	0.25
2-pentene (β-amylene)	36.07 (trans) ; 36.94 (cis)	12.58
piperylenes	42.0 (trans) ; 44.0 (cis)	0.65
isoamylenes	31.16 (2-methyl-1-butene) ; 38.57 (2-methyl-2-butene) ; 20.1 (3-methyl -1-butene)	7.26 12.06 1.34
cyclopentene	44.24	2.33
C5+ hydrocarbons	- - -	2.16
Specific weight (d_{15}^{15})	- - -	6.652

Table 9. Composition of C5 fractions and boiling points of main C5 hydrocarbons

14. Analysis of natural products

Gas chromatography coupled with MS represents a fast and cheap method for the separation and identification of compounds from natural product mixtures. Alkylation of animal fats and vegetable oils for biofuel fabrication is one of the industrial processes that needs analysis of natural products, both for raw materials and for products (Christie, 1989).

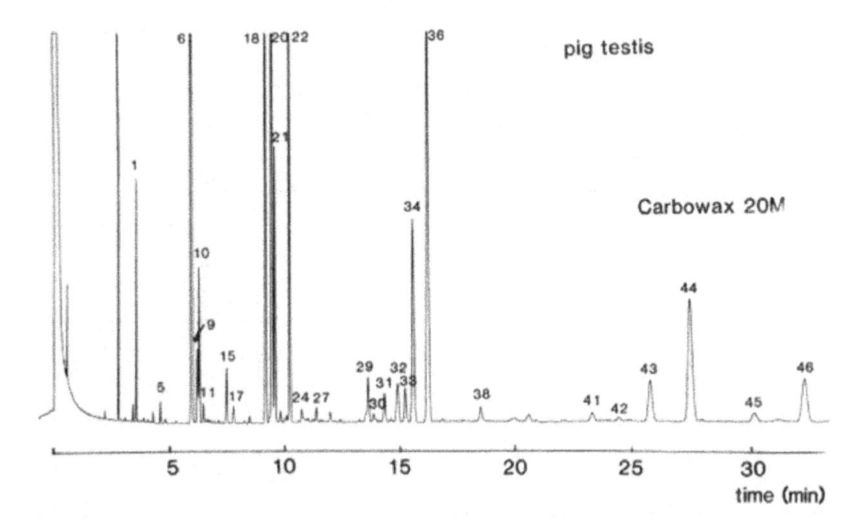

Fig. 3. Separation of the fatty acids (methyl esters) of pig testis on a fused silica column (25 m x 0.22 mm) coated with Carbowax 20M™ (Chrompak UK).

The separation of the methyl esters of the fatty acids of pig testis lipids on the Carbowax 20M™ column is illustrated in Figure 3. A gas chromatograph Carlo Erba Model 4130 split/splitless injection system was used. The carrier gas was hydrogen (1 ml/min), programme temperature was 165°C for 3 min and then temperature was raised at a rate of 4°C/min to 195°C and maintained for 23 min.

The three 16:1 isomers are well separated and distinct from the C_{17} fatty acids. The C_{18} components are separated from a minor 19:1 fatty acid and C_{20} unsaturated constituents. The 20:3(n-3), which co-chromatographs with 20:4(n-6) are just separable on a slightly more polar Silar 5CP™ column. C_{22}, important fatty acids are cleanly separated (Christie, 1989). Peaks can be identified in Table 10 (Christie, 1989).

Equivalent chain-lengths of the methyl ester derivatives of some fatty acids					
No	Fatty acid	Silicone	Carbowax	Silar 5CP	CP-Sil 84
1.	14.00	14.00	14.00	14.00	14.00
2	14-isobr	14.64	14.52	14.52	14.51
3	14-anteiso	14.71	14.68	14.68	14.70
4	14:1(n-5)	13.88	14.37	14.49	14.72
5	15:0	15.00	15.00	15.00	15.0
6	16:0	16.00	16.00	16.00	16.00
7	16-isobr	16.65	16.51	16.51	16.50
8	16-anteiso	16.73	16.68	16.68	16.69
9	16:1(n-9)	15.76	16.18	16.30	16.48
10	16:1(n-7)	15.83	16.25	16.38	16.60
11	16:1(n-5)	15.92	16.37	16.48	16.70
12	16:2(n-4)	15.83	16.78	16.98	17.47
13	16:3(n-3)	15.69	17.09	17.31	18.06
14	16:4(n-3)	15.64	17.62	17.77	18.82
15	17:0	17.00	17.00	17.00	17.00
16	17:1(n-9)	16.76	17.20	17.33	17.50
17	17:1(n-8)	16.75	17.19	17.33	17.51
18	18:0	18.00	18.00	18.00	18.00
19	18:1(n-11)	17.72	18.14	18.24	18.40
20	18:1(n-9)	17.73	18.16	18.30	18.47
21	18:1(n-7)	17.78	18.23	18.36	18.54
22	18:2(n-6)	17.65	18.58	18.80	19.20
23	18:2(n-4)	17.81	18.79	18.98	19.41
24	18:3(n-6)	17.49	18.85	19.30	19.72
25	18:3(n-3)	17.72	19.18	19.41	20.07
26	18:4(n-3)	17.55	19.45	19.68	20.59
27	19:1(n-8)	18.74	19.18	19.32	19.47
28	20:1(n-11)	19.67	20.08	20.22	20.35
29	20:1(n-9)	19.71	20.14	20.27	20.41
30	20:1(n-7)	19.77	20.22	20.36	20.50

Equivalent chain-lengths of the methyl ester derivatives of some fatty acids					
31	20:2(n-9)	19.51	20.38	20.58	20.92
32	20:2(n-6)	19.64	20.56	20.78	21.12
33	20:3(n-9)	19.24	20.66	20.92	21.43
34	20:3(n-6)	19.43	20.78	21.05	21.61
35	20:3(n-3)	19.71	20.95	21.22	21.97
36	20:4(n-6)	19.23	20.96	21.19	21.94
37	20:4(n-3)	19.47	21.37	21.64	22.45
38	20:5(n-3)	19.27	21.55	21.80	22.80
39	22:1(n-11)	21.61	22.04	22.16	22.30
40	22:1(n-9)	21.66	22.11	22.23	22.36
42	22:3(n-6)	21.40	22.71	22.99	23.47
43	22:4(n-6)	21.14	22.90	23.21	23.90
44	22:5(n-6)	20.99	23.15	23.35	24.19
45	22:5(n-3)	21.18	23.50	23.92	24.75
46	22:6(n-3)	21.04	23.74	24.07	25.07

*This document is part of the book "*Gas Chromatography and Lipids*" by William W. Christie and published in 1989 by P.J. Barnes & Associates (The Oily Press Ltd), who retain the copyright

Table 10. Equivalent chain-lengths of the methyl ester derivatives of some natural fatty acids*

15. Conclusions

1. Gas chromatography is an old and still current analytical method used in research and applied fields.
2. Separation of components in the sample is based on the difference of partition (adsorption or absorption) between a mobile phase in continuous motion, a permanent gas and a fixed phase called stationary phase, solid powder (separation by adsorption) or non-volatile liquid for separation by absorption.
3. Stationary phase is deposited in chromatography column, which is the heart of the gas chromatography apparatus.
4. Chromatography columns can be made of different materials: stainless steels, copper alloy, nickel, aluminium, plastics, glass, silica etc.
5. Chromatography columns can be packed, micro-packed, capillary, with differences in diameter, length, thickness of the packed-bed and, ultimately, with different column performance.
6. The materials most used for columns are glasses and silica, especially for capillary columns.
7. Using capillary columns have led to a considerable enlargement of the column length, shortening analysis time, increasing separation performance, the finding of stationary phases with improved performances both in terms of separability as well temperature resistance.
8. Efforts have increased to find the best stationary phase, allowing a good separation of geometric and optical isomers, for example polysiloxanes, and mixtures of different types and proportions with improved thermal resistance.

16. References

Avgul, N. N., Kiselev, A. V. & Poshkus, D. P. (1975). *Adsorbtsiya Gazovi Parovna Adnorodnikh Poverchnostyakh*, Khimiya, Moscow

Ciucanu, I. (1990). *Cromatografia de gaze cu coloane capilare*, Editura Academiei Romane, Bucharest

Christie, W. W. (1989). *Gas Chromatography and Lipids. A Practical Guide*, The Oily Press, Dundee, Scotland. URL: *www.lipidlibrary.aocs.org*

Comanescu, I. & Filotti, L. (2010). Vapour-liquid Equilibrium Data for C5 Saturated-Unsaturated Hydrocarbons Solutions with Solvents Monopropylene Glycol, Dipropylene Glycol or N-Methyl-2-Pyrrolidone, *Revista de Chimie* Vol.61: (10): 986-991

Grob, R. L. & Barry, E. F. (2004). *Modern Practice of Gas Chromatography*, Wiley-Interscience, New-York Jennings, W. G. (1986). *Comparisons of Fused Silica and Other Glasses Columns in Gas Chromatography*, Alfred Heuthig, Publishers, Heidelberg, Germany, 12-21

Jennings, W. G. (1997). *Analytical Gas Chromatography*, 2nd ed., Academic Press, San Diego, CA, 36-45

Kiselev, A. K. & Yashin, Ya. I. (1985). *Gas- and Flussigkeits-Adsorptionchromatographie*,VEB Deutscher Verlag der Wissenschaft, Berlin

Leary, J. J., Justice, J. B., Tsuge, S., Lowry, S. R. & Isenhour, T. L. (1973). Correlating gas chromatographic liquid phases by means of nearest neighbor technique: a proposed set of twelve preferred phases, *Journal of Chromatography Science* Vol.(11): 201-206

Leibnitz, E. & Struppe, H. G. (1984). Ed. *Handbuch der Gas-Chromatographie Vol. 2*, Akademische Verlagesellschaft Geest & Portig, Leipzig

Poole, C. F. (2003). *The Essence of Cromatography*, Elsevier, Amsterdam

Rotzsche, H. (1991). *Stationary Phases in Gas Chromatography*, Elsevier, Amsterdam

4

Derivatization Reactions and Reagents for Gas Chromatography Analysis

Francis Orata
Masinde Muliro University of Science and Technology, Kenya

1. Introduction

Derivatization reactions are meant to transform an analyte for detectability in Gas Chromatography (GC) or other instrumental analytical methods. Derivatization in GC analysis can be defined as a procedural technique that primarily modifies an analyte's functionality in order to enable chromatographic separations. A modified analyte in this case will be the product, which is known as the derivative. The derivative may have similar or closely related structure, but not the same as the original non-modified chemical compound.

Volatility of sample is a requirement for GC analysis. Derivatization will render highly polar materials to be sufficiently volatile so that they can be eluted at reasonable temperatures without thermal decomposition (Knapp, 1979) or molecular re-arrangement (Kühnel et al., 2007; Blau and King 1979). Understanding the chemistry of the analytes, derivatizing reagents used in sample preparation, and the detailed functionality of Gas Chromatography are important to get reliable results. For GC analysis, compounds containing functional groups with active hydrogens such as -SH, -OH, -NH and -COOH are of primary concern because of the tendency of these functional groups to form intermolecular hydrogen bonds (Zaikin and Halket, 2003). These intermolecular hydrogen bonds affect the inherent volatility of compounds containing them, their tendency to interact with column packing materials and their thermal stability (Sobolevsky et al., 2003). Since GC is used to separate volatile organic compounds, modification of the functional group of a molecule by derivatization enables the analysis of compounds that otherwise can not be readily monitored by GC. Derivatization process either increases or decreases the volatility of the compound of interest. It also reduces analyte adsorption in the GC system and improves detector response, peak separations and peak symmetry.

In addition to particular analytes such as pharmaceuticals, biomolecules such us organic acids, amides, poly-hydroxy compounds, amino acids, pesticides and other persistent organic compounds, new classes of compounds of interest for example fluorinated alkylated substances and polycyclic aromatic hydrocarbons continue to emerge. It is necessary to develop and/or improve on chemical analytical methods and hence the need to familiarize with derivatization methods that are applicable to GG analysis. Generally derivatization is aimed at improving on the following aspects in Gas Chromatography.

i. Suitability

Suitability is the form of compounds that is amenable to the analytical technique. For GC, it is a requirement that the compound to be analyzed should be volatile with respect to gas chromatographic analysis conditions, as compared to liquid chromatography (LC), where the compound of interest should be soluble in the mobile phase. Therefore, derivatization procedure modifies the chemical structure of the compounds so that they can be analyzed by the desired technique.

ii. Efficiency

Efficiency is the ability of the compound of interest to produce good peak resolution and symmetry for easy identification and practicability in GC analysis. Interactions between the compounds themselves or between the compounds and the GC column may reduce the separation efficiency of many compounds s and mixtures (Knapp, 1979). Derivatization of analyte molecules can reduce these interactions that interfere with analysis. Also, compounds that co-elute or have poor resolution from other sample components during separation in GC can frequently be resolved by an appropriate derivative.

iii. Detectability

Detectability is the outcome signal that emanates from the interaction between the analyte and the GC detector. Increasing the amounts of materials will impact the range at which they can be detected in Gas chromatography. This can be achieved either by increasing the bulk of the compound or by introducing onto the analyte compound, atoms or functional groups that interact strongly with the detector and hence improve signal identification. For example the addition of halogen atoms to analyte molecules for electron capture detectors (ECD) and the formation of trimethylsilyl (TMS) ether derivatives to produce readily identifiable fragmentation patterns and mass ions (Knapp, 1979).

1.1 Derivatization reagent

Derivatization reagent is the substance that is used to chemically modify a compound to produce a new compound which has properties that are suitable for analysis in GC or LC. The following criteria must be used as guidelines in choosing a suitable derivatization reagent for GC analysis.

i. The reagent should produce more than 95 % complete derivatives.
ii. It should not cause any rearrangements or structural alterations of compounds during formation of the derivative.
iii. It should not contribute to loss of the sample during the reaction.
iv. It should produce a derivative that will not interact with the GC column.
v. It should produce a derivative that is stable with respect to time.

1.2 Objectives for derivatization

The following outlined objectives among others can be achieved by application of proper derivatization procedures;

i. Improvement of resolution and reduce tailing of polar compounds which may contain -OH, -COOH, =NH, $-NH_2$, -SH, and other functional groups.

ii. Analysis of relatively nonvolatile compounds.
iii. Reduction of volatility of compounds prior to GC analysis.
iv. Improvement of analytical efficiency and hence increase detectability.
v. Stabilization of compounds for GC analysis.

2. Types of derivatization reactions

Derivatization reactions used for gas chromatography (GC) fall into three general reaction types namely; Alkylation of which the general process is esterification, Acylation and Silylation. Through these three processes, highly polar materials such as organic acids, amides, poly-hydroxy compounds, amino acids are rendered suitable for GC analysis by making them sufficiently volatile. These general processes are discussed below.

2.1 Alkylation

Alkylation is mostly used as the first step for further derivatizations or as a method of protection of certain active hydrogens in a sample molecule. It represents the replacement of active hydrogen by an aliphatic or aliphatic-aromatic (e.g., benzyl) group in process referred to as esterification. Equation 1 below shows the general reaction equation representing the esterification process.

$$RCOOH + PhCH_2X \rightarrow RCOOCH_2Ph + HX$$

Equation 1: General reaction for esterification process: X = halogen or alkyl group R, H = another alkyl group R.

The principal chromatographic use of this reaction is the conversion of organic acids into esters, especially methyl esters that produce better chromatograms than the free acids. Alkylation reactions can also be used to prepare ethers, thioethers and thioesters, N-alkylamines, amides and sulphonamides (Danielson, 2000). In general, the products of alkylation are less polar than the starting materials because active hydrogen has been replaced by an alkyl group. The alkyl esters formed offer excellent stability and can be isolated and stored for extended periods if necessary. In esterification an acid reacts with an alcohol to form an ester. In the reaction, a catalyst more often an inorganic acid such as hydrochloric acid or thionyl chloride (Zenkevich, 2009) is recommended for example, in the trans-esterification of fats or oils (Sobolevsky et al., 2003).

2.1.1 Derivatization reagents used in alkylation

Common derivatization reagents for the Alkylation type of reactions are Dialkylacetals, Diazoalkales, Pentafluorobenzyl bromide (PFBBr), Benzylbromide, Boron trifluoride (BF_3) in methanol or butanol and Tetrabutylammonium hydroxide (TBH) among others. Alkylation reagents can be used alone to form esters, ethers and amides or they can be used in conjunction with acylation or silylation reagents. The reaction conditions can vary from strongly acidic to strongly basic with both generating stable derivatives. However, it must be noted that the reagents are more limited to amines and acidic hydroxyls and that the reaction conditions are frequently severe while the reagents are often toxic. Some derivatization reagents and their respective derivatization procedures in alkylation reactions are discussed below.

2.1.1.1 Dialkylacetals

Dimethylformamide (DMF) is an example of dialkylacetals with a general formula $CH_3CH_3NCHOROR$ are used to esterify acids to their methyl esters. Dialkylacetals have a wider applicability for the derivatization of a number of functional groups containing reactive hydrogens. Because the principal reaction product is dialkylacetals (DMF), the isolation of the derivative is not required and the reaction mixture can be injected directly into the gas chromatograph (Regis, 1999). This reagent is an excellent first choice for derivatization of a compound for which there is no published method available. The reaction between N, N-dimethylformamide dimethylacetal and Carboxylic acid is as follows (Equation 2).

$$CH_3CH_3NCHOROR + R`COOH \rightarrow R`COOR + ROH + CH_3CH_3NCHO$$

Equation 2: The reaction between N, N-dimethylformamide dimethylacetal and Carboxylic acid.

Although carboxylic acids, phenols, and thiols react quickly with DMF, to give the corresponding alkyl derivatives, hydroxyl groups are not readily methylated. During derivatization procedure, care should be taken because *N, N*-dimethylformamide dimethylacetals are moisture sensitive. The reagents work quickly with derivatization taking place just upon dissolution. The reaction is suitable for flash alkylation, where derivatization takes place in the injection port. Since adjustment of polarity and volatility of the sample is possible, this can allow change of retention time. It is worth noting that the reagents will react with water to give the corresponding alcohol though traces of water will not affect the reaction as long as you have an excess of the acid.

A rapid procedure for the derivatization of both carboxylic and amino acids using DMF-Dialkylacetal reagents is by dissolving the sample in 0.5 mL of a 1:1 solvent of choice/ reagent mixture by heating to 100 °C. For example, it is advisable to use pyridine for fatty acids, acetonitrile for amino acids as solvent of choice (Thenot and Horning, (1972).

Alternative method comprises of combining 50 mg fatty acid and 1 ml DMF in a reaction vial. Cap the vial and heat at 60 °C for 10 - 15 minutes or until dissolution is complete and analyze the cooled sample in gas chromatography (Thenot et al., 1972).

2.1.1.2 Diazoalkales

The main reagent in the diazoalkales group is diazomethane. Diazomethane (N_2CH_2) is the quickest and cleanest method available for the preparation of analytical quantities of methyl esters. The reaction of diazomethane with a carboxylic acid is quantitative and essentially instantaneous in ether solutions. In the presence of a small amount of methanol as catalyst, diazomethane reacts rapidly with fatty acids, forming methyl esters. Elimination of gaseous nitrogen drives the reaction forward. The reaction for the conversion of carboxylic acids to methyl esters (Equation 3) is outlined below:

$$RCOOH + {}^{-}CH_2N^{+}N \rightarrow RCOOCH_3 + N_2$$

Equation 3: The reaction for the conversion of carboxylic acids to methyl esters

The yield is high and side reactions are minimal. Sample sizes of 100 - 500 µl are easily derivatized and the isolation of the methyl esters is simple and quantitative when dealing with acids having chain lengths from C8 to C24. However, care should be taken in handling diazomethane because it is carcinogenic, highly toxic, and potentially explosive.

2.1.1.3 Pentafluorobenzyl bromide (PFBBr) and Pentafluorobenzyl-hydroxylamine hydrochloride (PFBHA)

Pentafluorobenzyl bromide ($C_7H_2F_5Br$) and also Pentafluorobenzyl-hydroxylamine hydrochloride can be used to esterify phenols, thiols, and carboxylic acids. Equation 4 below is an example for PFBBr derivatization process.

$$R`OR + C_7H_2F_5Br \rightarrow R`OC_7H_2F_5 + RBr$$

Equation 4: Reaction of PFBBr with either phenols, thiols or carboxylic acids: R = Hydrogen

Derivatization can be done with O-(2,3,4,5,6-pentafluorobenzyl)-hydroxylamine hydrochloride (PFBHA) for carbonyls and pentafluorobenzyl bromide (PFBBr) for carboxylic acid and phenol groups for gas (GC/MS) in an electron impact mode (EI) and a gas chromatograph/ion trap mass spectrometry (GC/ITMS) in both chemical impact and EI modes. To confirm different isomers, the PFBHA-derivatives of analytes can be rederivatized by silylation using N, O-bis (trimethylsilyl)-trifluoroacetamide (BSTFA) (Jang and Kamens, 2001).

Example for the application of this method is from the work of Allmyr et al., (2006) where the analysis of Triclosan in human plasma and milk was accomplished by the conversion of Triclosan into its pentafluorobenzyl ester by adding 2 ml of H_2O (Milli-Q), 50 µl of 2M KOH, 10 µl pentafluorobenzyl chloride in 10% toluene and 0.3 g NaCl (more sodium chloride was added if emulsion formed upon mixing) to the sample extract and shaking the tube vigorously for 2 minutes. This was followed by extraction of the aqueous phase with 2 ml of n-hexane. Then, 3 ml of 98 % H_2SO_4 was added to the extract, and the tube inverted 60 times and another extraction using 2 ml of n-hexane followed. The final extract was reduced to 2 ml under a gentle stream of nitrogen gas at room temperature. Approximately 0.5 ml of the extract was injected to GC for analysis.

i. Organic Acids derivatisation

The following procedure by Chien et al., (1998) and Galceran et al., (1995) is a variation of methods previously utilized for organic acid and phenol analysis. In the method, sample extracts for organic acid/phenol derivatization are first dried by passing a gentle stream of nitrogen in the sample. Once the solvent has been evaporated, acetone is added to bring each sample to a volume of 500 µL. Each sample is added 20 µL of 10% PFBBr solution and 50 µL of 1,4,7,10,13,16-hexaoxacyclooctadecane $[C_2H_4O]_6$ (18-crown-6 ether solution), both in acetone. Approximately 10 mg of potassium is added to each extract, and the extracts are capped and sonicated for three hours. Upon completion of sonication the acetone is evaporated by passing a gentle stream of nitrogen, and the residue dissolved into hexane for GC analysis.

Equation 5 below shows the chemical reactions for the conversion of organic acids and phenols into their pentafluorobenzyl esters and ether respectively using pentafluorobenzyl bromide (PFBBr) derivatization.

i. $$ROCOH + C_7H_2F_5Br \xrightarrow[18-c-6\ ether]{K_2CO_3} C_7H_2F_5OCOR + HBr$$

ii. $PhOH + C_7H_2F_5Br \xrightarrow[18-c-6\ ether]{K_2CO_3} C_7H_2F_5OPh + HBr$

Equation 5: Reactions for the conversion of organic acids and phenols into their pentafluorobenzyl esters (i) and ether (ii) respectively using pentafluorobenzyl bromide (PFBBr).

In another example of pentafluorobenzylation procedure (Naritsin and Markey, 1996 & Kawahara, 1968), 20 μl of PFBBr is added to the vial containing already extracted and cleaned sample in methyl chloride. The vial is then capped and shaken for 20 - 30 minutes. Gas chromatography analysis can then be accomplished using either of the two methods depending on the available detector. A portion of methylene chloride phase can be injected into the chromatograph for FID analysis or it can be evaporated using a gentle stream of nitrogen gas and re-dissolve in benzene for ECD analysis. Reagents containing fluorinated benzoyl groups are good for ECD analysis.

ii. Carbonyls

The derivatization process performed by Spaulding et al., (2003), Destaillats et al., (2001) & Rao et al., (2001) for the analysis of carbonyl involves first, the reduction in volume of sample extracts to < 50 μL under a gentle stream of nitrogen. Once the extract volume has been reduced, a 9:1 (v/v) mixture of carbonyl-free Acetonitrile: Dichloromethane (DCM) is added to bring the sample to a volume of 500 μL. This is followed by the addition of 50 $mgmL^{-1}$ solution of *O*-(2, 3, 4, 5, 6-pentafluorobenzyl) hydroxylamine hydrochloride (PFBHA) in methanol that will result to a target PFBHA concentration of 5 mM. The sample is left at room temperature for a period of 24 hours then subsequently analyzed using GC. The balanced chemical reaction where *O*-(2, 3, 4, 5, 6-pentafluorobenzyl) hydroxylamine hydrochloride (PFBHA) is used for the conversion of carbonyls into their pentafluorobenzyl oximes is provided below (Equation 6).

$$R'OCR'' + C_6F_5\text{-}ONH_2 \rightarrow C_6F_5\text{-}ONR'CR'' + Isomer(s) + H_2O$$

Equation 6: Chemical reaction for the conversion of carbonyls into pentafluorobenzyl oximes using *O*-(2, 3, 4, 5, 6-pentafluorobenzyl) hydroxylamine hydrochloride (PFBHA).

The PFBHA derivatization process for most of the carbonyl compounds could be completed in 2 h at room temperature (Bao et al., 1998).

2.1.1.4 Benzylbromide

Benzyl bromide reacts with the acid part of an alkyl acid to form an ester, and therefore increase the volatility of the analyte of interest. Orata et al., (2009) determined long chain perfluorinated acids namely; perfluoro-n-octanoic acid (PFOA), 2H-perfluoro-2-octenoic acid (FHUEA), 2H-perfluoro-2-decenoic acid (FOUEA) and 2H-perfluoro-2-dodecenoic acid (FNUEA) in biota (fish) and abiota (water) by derivatization and subsequent analysis by GC/MS. The method involved derivatization of long chain perfluorinated compounds using benzyl bromide solution and acetone to form benzylperfluorooctanoate (benzyl ester) as presented by the equation 7 below which shows the modification of perfluorooctanoic acid to the respective ester.

$$CF_3(CF_2)_6COOH + C_7H_7Br \rightarrow C_{15}H_7F_{15}O_2 + HBr$$

Equation 7: Convertion of Perfluorooctanoic acid to Benzylperfluorooctanoate using benzyl bromide.

Long Chain Perfluorinated alkyl acids derivatization procedure by Orata et al (2009) is as follows: Following extraction and sample clean up procedure, the sample is dried by a gentle stream of nitrogen. The sample residue is then dissolved in 0.5 mL of acetone and 0.1 mL of benzyl bromide solution is added to it. The tube is heated at 80 °C for 15 min, cooled, and evaporated to dryness with nitrogen. The residue is then dissolved in 1 mL of methylene chloride for GC analysis. Results obtained in the study (Orata et al, 2009) demonstrate that GC/MS can be an alternative to the LC/MS method for quantification of perfluorinated acids in contaminated areas, where expected higher concentration of the analyte is expected. It was in the view of the author (Orata et al, 2009) that the method can be further improved to lower the detection limits.

2.1.1.5 Tetrabutylammonium hydroxide (TBH)

Derivatization of a carboxylic acid with tetrabutylammonium hydroxide (TBH) forms butyl ester, which will allow a longer retention times in a GC column (Lin et al., 2008). The reagent is most commonly used for low molecular weight acids and is especially suitable for low molecular weight amines (Regis, 1999). Equation 8 represents the derivatization reaction for the conversion of carboxylic acid to alkyl esters using TBH.

$$[N(CH_3)_4]^+[OH]^- + RCOOH \rightarrow RCOOC_4H_9$$

Equation 8: Conversion of carboxylic acid to alkyl esters using TBH.

The following derivatization procedure can be used for flash alkylation which is suitable for biological fluids and thermally stable fatty acids analysis. Biological fluid or tissue is first extracted using toluene. Then 4 mL of the extract is transferred into a nipple tube and evaporate under a gentle stream of nitrogen gas at 60 °C. 25 µL of TBH solution is added to dissolve the residue. After 30 minutes, 4 µL of the dissolved residue is injected directly onto the chromatograph. The injection port temperature is set at 260 °C or above in GC instrument.

2.1.1.6 Boron trifluoride (BF3) in methanol or butanol

Boron trifluoride (BF_3) in methanol or *n*-Butanol has a general formula is F_3B: $HO\text{-}C_nH_{2n+1}$ where (n = 1 or 4). This reagent is convenient and inexpensive method for forming esters. It is most commonly used to form methyl (butyl) ester by reacting it with acids, as shown by the following general equation 9.

$$F_3B{:}\ HO\text{-}C_nH_{2n+1} + RCOOH \rightarrow RCOOC_nH_{2n+1}$$

Equation 9: General equation for formation of alkyl esters using Boron trifluoride (BF_3) as a derivatization reagent.

Boron trifluoride reagent forms fluoroboron compounds by reaction with atmospheric oxygen and methanol and therefore making it prone to instability (Christie 1993). Therefore, it should always be stored at the refrigerator temperature discarded after a few months use.

In a study by Lough (1964) were polyunsaturated fatty acids were analyzed, it was observed that boron trifluoride-methanol will cleave the rings in cyclopropane fatty acids (commonly encountered in microorganisms), and it causes cis-trans isomerization of double bonds in conjugated fatty acids. In addition, it reacts with the antioxidant Butylated hydroxytolune (BHT) to produce spurious peaks in chromatograms.

The following esterification procedure can be used in derivatization using Boron trifluoride. To a 100 mg organic acid in a vial, 3 mL BF_3 in methanol or BF_3 in n-butanol is added and heated to 60 °C for 5 to 10 minutes. Heating temperature and time may vary i.e. Ribeiro et al., (2009) used 90 °C for 10 minutes for the process. This is followed by cooling and transferring the mixture to a separating funnel with 25 mL hexane. Further, the sample is washed 2 times with saturated NaCl solution, and then dried over anhydrous Na_2SO_4. The solvent is evaporated to concentrate the sample and lastly injected onto column for GC analysis.

2.2 Silylation

Silylation is the most prevalent derivatization method as it readily volatizes the sample and therefore very suitable for non-volatile samples for GC analysis. Silylation is the introduction of a silyl group into a molecule, usually in substitution for active hydrogen such as dimethylsilyl [$SiH(CH_3)_2$], t-butyldimethylsilyl [Si $(CH_3)_2C(CH_3)_3$] and chloromethyldimethylsilyl [$SiCH_2Cl(CH_3)_2$]. Replacement of active hydrogen by a silyl group reduces the polarity of the compound and reduces hydrogen bonding (Pierce, 1968). Many hydroxyl and amino compounds regarded as nonvolatile or unstable at 200 – 300 °C have been successfully analyzed in GC after silylation (Lin et al., 2008 & Chen et al., 2007). The silylated derivatives are more volatile and more stable and thus yielding narrow and symmetrical peaks (Kataoka, 2005).

2.2.1 Silylation reaction and mechanism

The silylation reaction is driven by a good leaving group, which means a leaving group with a low basicity, ability to stabilize a negative charge in the transitional state, and little or no back bonding between the leaving group and silicon atom (Knapp, 1979). The mechanism involves the replacement of the active hydrogens (in -OH, -COOH, -NH, -NH_2, and -SH groups) with a trimethylsilyl group. Silylation then occurs through nucleophilic attack (SN_2), where the better the leaving group, the better the siliylation. This results to the production of a bimolecular transition state (Kühnel et al., 2007) in the intermediate step of reaction mechanism. The general reaction for the formation of trialkylsilyl derivatives is shown by equation 10. The leaving group in the case of trimethylchlorosilane (TMCS) is the Cl atom.

$$\text{Sample}-\ddot{\text{O}}(\text{H}): + \text{CH}_3-\text{Si}(\text{CH}_3)_2-\text{X} \longrightarrow \left[\text{Sample}-\ddot{\text{O}}(\text{H})\overset{\delta+}{-}\text{Si}(\text{CH}_3)_3\overset{\delta-}{-}\text{X}\right] \longrightarrow \text{Sample}-\text{O}-\text{Si}(\text{CH}_3)_2-\text{CH}_3 + \text{HX}$$

Equation 10: General reaction mechanism for the formation of trialkylsilyl derivatives for trimethylchlorosilane, X = Cl

In silylation derivatisation, care must be taken to ensure that both sample and solvents are dry. Silyl reagents generally are moisture sensitive, and should be stored in tightly sealed containers (Sobolevsky et al., 2003) and therefore the solvents used should be as pure and as little as possible. This will eliminate excessive peaks and prevent a large solvent peak. In silylation, pyridine is the most commonly used solvent. Although pyridine may produce peak tailing, it is an acid scavenger and will drive the reaction forward. In many cases, the need for a solvent is eliminated with silylating reagents. The completion of the derivatization process in silylation is usually observed when a sample readily dissolves in the reagent. According to Regis (1999) the ease of reactivity of the functional group toward silylation follows the order:

Alcohol > Phenol > Carboxyl > Amine > Amide /hydroxyl

For alcohols, the order will be as follows:

Primary > Secondary > Tertiary

Many reagents will require heating that is not in excess of 60 °C for about 10 - 15 minutes, to prevent breakdown of the derivative. Although hindered products may require long term heating.

2.2.2 Derivatization reagents used in Silylation

Reagents used for the silylation derivatization process include Hexamethyldisilzane (HMDS), Trimethylchlorosilane (TMCS), Trimethylsilylimidazole (TMSI), Bistrimethylsilylacetamide (BSA), Bistrimethylsilyltrifluoroacetamide (BSTFA), N-methyl-trimethylsilyltrifluoroacetamide (MSTFA), Trimethylsilyldiethylamine (TMS-DEA), N-methyl-N-t-butyldimethylsilyltrifluoroacetamide (MTBSTFA), and Halo-methylsilyl derivatization reagents. Halo-methylsilyl derivatization reagents which although not discussed in this chapter can produce both silylated and halogenated derivatives for ECD. Silyl reagents will react with both alcohols and acids to form trimethylsilyl ethers and trimethylsilyl esters respectively. These derivatives formed are volatile and for the most part, are easily separated (Scott, 2003). Silyl reagents are compatible with most detection systems but, if they are used in excess, they can cause difficulties with flame ionization detectors (FID) (Sobolevsky et al, 2003). Silyl reagents are influenced by both the solvent system and the addition of a catalyst. A catalyst (e.g., trimethylchlorosilane or pyridine) increases the reactivity of the reagent.

Reagents that introduce a t-butyldimethylsilyl group in place of the trimethylsilyl group were developed to impart greater hydrolytic stability to the derivatives. These t-butyldimethylsilyl derivatives not only have improved stability against hydrolysis, but they also have the added advantage of distinctive fragmentation patterns, which makes them useful in a variety of GC/MS applications. Most trimethylsilyl and t-butyldimethylsilyl derivatives have excellent thermal stability and are amenable to a wide range of injection and column conditions (Pierce 2004). Silylation gives ability to derivatize a wide variety of compounds for GC analysis. In addition a large number of silylating reagents available.

2.2.2.1 Bis(trimethylsilyl)-acetamide (BSA)

Bis(trimethylsilyl)-acetamide (BSA) was the first widely used silylating reagent. The strength of BSA as a strong silylating reagent is enhanced more because acetamide is a good leaving group. BSA reacts under mild conditions and produces relatively stable by-products. However, the by-product which is trimethylsilyl -acetamide sometimes produces peaks that overlap those of other volatile derivatives as shown in the equation 11 below.

$$H_3C-C(O-TMS)=N-TMS + H-Y-R \longrightarrow TMS-Y-R + H_3C-C(=O)-N(H)-TMS$$

Equation 11: Equation showing the by-product which is trimethylsilyl -acetamide from Bis(trimethylsilyl)-acetamide silylation reagent: TMS = $Si(CH_3)_3$, Y = O, S, NH, NR\`, COO, R, R\` = Alk, Ar.

BSA forms highly stable trimethylsilyl derivatives with most organic functional groups under mild reaction conditions but their mixtures also oxidize to form silicon dioxide, which can foul FID detectors. Experimentally, BSA has been found to be very efficient and only a small amount of reagent is required. Moreso the derivatization process takes a shorter time at room temperature. Saraji and Mirmahdieh, (2009) studied the influence of different parameters on the in-syringe derivatization process using BSA were three parameters, the amount of the BSA, the reaction time and the reaction temperature were investigated to achieve the highest derivatization reaction yield. The results showed that 0.2 μL volume of BSA derivatization reagent is enough to obtain the maximum derivatization reaction yield for all compounds in 2 min time at room temperature.

2.2.2.2 Bis(trimethylsilyl)trifluoroacetamide (BSTFA)

N, N-bis(trimethyl-silyl)trifluoro-acetamide (BSTFA) just like bis(trimethylsilyl)-acetamide (BSA) are the two most popular reagents for Silylation type of derivatization. They react rapidly with organic acids to give high yields. Gerhke, (1968) developed a method in which a few milligram of an acid is placed in a vial and about 50 μl of BSA or BSTFA is added to it. The reaction can be expected to be complete directly within the solution, but the mixture can also be heated for 5 to 10 min. at 60 °C to ensure that reaction is really complete, before GC analysis. BSTFA reacts faster (as shown using the reaction 12 below) and more completely than BSA, due to presence of trifluoroacetyl group.

$$F_3C-C(O-TMS)=N-TMS + H-Y-R \longrightarrow TMS-Y-R + F_3C-C(=O)-N(H)-TMS$$

Equation 12: Silylation reaction using N, N-bis(tri methyl-silyl)trifluoro-acetamide: TMS = $Si(CH_3)_3$, Y = O, S, NH, NR\`, COO, R, R\` = Alk, Ar.

The high volatility of BSTFA and its byproducts results in separation of early eluting peaks. In addition, its stable products result in low detector noise and fouling. Addition of trimethylchlorosilane (TMCS) catalyzes reactions of hindered functional groups in secondary alcohols and amines.

In the method used by Schauer et al., (2002), sample extracts for analysis of hydroxylated polycyclic aromatic hydrocarbons (Hydroxy-PAHs), the sample extracts are first reduced in volume to 200 μL under a gentle stream of nitrogen. To each sample 20 μL of freshly prepared solution of 10% (v/v) Trimethylchlorosilane (TMCS) in N-O-bis (trimethylsilyl)-trifluoroacetamide (BSTFA) is added. The samples are capped, wrapped with Teflon tape, and heated at 45 °C for 24 hours to convert the targeted analytes to their trimethylsilyl derivatives. The balanced reaction (Equation 13) for this conversion to trimethyl silyl (TMS) ethers using N-O-bis (trimethylsilyl)-trifluoroacetamide (BSTFA) is provided in below.

TMCS
Pyridine

Equation 13: Derivatization of Hydroxylated polycyclic aromatic hydrocarbons (Hydroxy-PAHs) into trimethyl silyl (TMS) ethers using BSTFA.

Caution must be exercised to prevent any water from entering the samples as this will lead to hydrolysis of the BSTFA reagent and prevent any of the targeted analytes from undergoing derivatization.

Szyrwińska et al., (2007) and Kuo and Ding, (2004) used N, O-bis-(trimethylsilyl)trifluoroacetamide (BSTFA) containing 1% trimethylchlorosilane (TMCS), and bromoacetonitrile (BAN) reagents to analyze standard solutions of Bisphenol-A (BPA) and extracts of powdered milk.

The method was by derivatization using Trimethylsilylation with the derivatizing reagent which was BSTFA + 1% TMCS. Bisphenol-A standard solution (200 μL) was placed in a vial (1 mL) and evaporated to dryness under a gentle stream of nitrogen at 60 °C. Silylating agent (BSTFA containing 1 % TMCS; 100 μL) was added to the residue and the vial was vortex mixed and heated at 80 °C for 30 min. After cooling, the derivatized solution was evaporated to dryness and the residue was re-dissolved in 100 μL chloroform. This solution (1 μL) was analyzed by GC-MS.

2.2.2.3 N-methyl-trimethylsilyltrifluoroacetamide (MSTFA)

N-methyl-trimethylsilyltrifluoroacetamide (MSTFA) is the most volatile of the trimethylsilyl acetamides. It is most useful for the analysis of volatile trace materials where the derivatives may be near the reagent or by-product peak. The general equation of the reaction (Equation 14) using MSTFA is shown below;

$$F_3C{-}C({=}O){-}N(TMS){-}CH_3 + H{-}Y{-}R \longrightarrow TMS{-}Y{-}R + F_3C{-}C({=}O){-}N(H){-}CH_3$$

Equation 14: Silylation reaction using N-methyl-trimethylsilyltrifluoroacetamide (MSTFA): TMS = $Si(CH_3)_3$, Y = O, S, NH, NR`, COO, R, R` = Alk, Ar.

Silylation derivatization procedures by Butts, (1972) using BSTFA, BSA, or MSTFA formed the basis for derivatisation application that was used to analyze over 200 organic compounds including carboxylic acids, amines, alcohols, phenols, and nucleic acids. The basis of the procedure is as follows: to a 1 - 5 mg sample, 100 µL each of pyridine as a solvent and silylating reagent are added. The mixture is capped and heated to 60 °C for 20 minutes. For moderately hindered or slowly reacting compounds, BSTFA is recommended for use with 1 or 10 % TMCS catalyst. Under extreme conditions compounds may require heating for up to 16 hours. When Ketones are derivatized using this procedure, they may form 15 - 20 % enol trimethylsilyl esters. The esters can be eliminated by first forming a methoxime. In the case of Amino acids the reaction may be required to be in a sealed tube or vial (Butts, 1972). The samples are heated cautiously, near the boiling point of the mixture until a clear solution is obtained.

2.2.2.4 Hexamethyldisilzane (HMDS)

Hexamethyldisilzane (HMDS) is a weak donor, as it has symmetry. If it is used for derivatization, it will attack only easily silylated hydroxyl groups. Equation 15 shows the reaction using hexamethyldisilzane.

$$(CH_3)_3Si{-}NH{-}Si(CH_3)_3 + H{-}Y{-}R \longrightarrow (CH_3)_3Si{-}Y{-}R + H_2N{-}Si(CH_3)_3$$

Equation 15: Silylation reaction using Hexamethyldisilzane (HMDS): Y = O, S, NH, NR`, COO, R, R` = Alk, Ar.

Hexamethyldisilzane is a weak TMS donor that can be used for silylation of carbohydrates. It can be used as mixture with pyridine and trifluoroacetic acid. The derivatization procedure is simple as shown in this procedure where 2 ml of hexamethyldisilazane, 2 ml of pyridine and 175 µL of trifluoroacetic acid are added to the samples. Silylation is then performed at 60 °C for 1 hour.

2.2.2.5 Trimethylchlorosilane (TMCS)

Trimethylchlorosilane (TMCS) is also a weak donor. In addition, it produces hydrochloric acid as a by product with is acidic. It is therefore not commonly used. However, it is often found as a catalyst to increase TMS donor potential. An example of derivatization reaction using Trimethylchlorosilane (TMCS) is shown in equation 16.

$$(CH_3)_3Si{-}Cl + H{-}Y{-}R \longrightarrow (CH_3)_3Si{-}Y{-}R$$

Equation 16: Silylation reaction using Trimethylchlorosilane (TMCS): Y = O, S, NH, NR`, COO, R, R` = Alk, Ar.

2.2.2.6 Trimethylsilylimidazole (TMSI)

Trimethylsilylimidazole (TMSI) is not a weak donor, but it is selective as it reacts with alcohols and phenols but not amines or amides (nitrogen groups). Since it is selective, it will target the hydroxyls in wet sugars and also derivatize the acid sites of amino acids. It will leave the amino group free for fluorinated derivatization. An example of reaction equation using TMSI is shown below (Equation 17).

$$(CH_3)_3Si{-}N(\text{imidazole}) + H{-}O{-}R \longrightarrow (CH_3)_3Si{-}O{-}R + H{-}N(\text{imidazole})$$

Equation 17: Silylation reaction using Trimethylsilylimidazole (TMSI): TMS = R, R` = Alk, Ar.

The derivatives produced are suitable for ECD analysis. Isoherranen and Soback, (2000) used the following derivatization procedure for the determination of Gentamicin. In the procedure, the sample solution was evaporated to dryness under a stream of nitrogen in an auto sampler vial and the dry residue was dissolved in 50 ml of anhydrous pyridine. To the pyridine solution, 100 ml of TMSI were added and the vial was closed and incubated for 15 min at 60 °C. Thereafter, 70 ml of Trifluoroacetic anhydride (TFAA) were added and the vial was closed and incubated for 60 min at 60 °C.

2.2.2.7 Trimethylsilyldiethylamine (TMS-DEA)

Trimethylsilyldiethylamine (TMS-DEA) reagent is used for derivatizing amino acids, antibiotics, urea-formaldehyde condensates, steroids and carboxylic acids and it also targets hindered compounds. Hydrolysis of TMS derivatives and reagents produces hexamethyldisiloxane $[(CH_3)_3SiOSi(CH_3)]$. Hexamethyldisiloxane is quite inert and does not interfere in the reaction or produce byproducts with the sample. The reaction by product diethylamine is very volatile and the reaction can be driven to completion by evaporating the diethylamine produced. Equation 18 shows the derivatisation reaction using TMS-DEA reagent.

$$\text{TMS-N }(C_2H_5)_2 + \text{H-Y-R} \rightarrow \text{TMS-Y-R} + \text{H-N }(C_2H_5)_2$$

Equation 18: Silylation reaction using Trimethylsilyldiethylamine (TMS-DEA) reagent: TMS = $Si(CH_3)_3$, Y = O, S, NH, NR`, COO, R, R` = Alk, Ar.

Because of its high volatility, it is eluted with the solvent or reagent and usually does not interfere with the chromatogram.

2.2.2.8 N-methyl-N-t-butyldimethylsilyltrifluoroacetamide (MTBSTFA)

Silylation derivatisaion using N-methyl-N-t-butyldimethylsilyltrifluoroacetamide (MTBSTFA) replaces the active hydrogen with tert-Butyldimethylsilyl (t-BDMS) group. The tert-Butyldimethylsilyl derivatives which are more resistant to hydrolysis and can be up to 10,000 times more stable than TMS derivatives. N-methyl-N-t-butyldimethylsilyltrifluoroacetamide will target sulfonic and phosphoric groups if present and it is suitable for GC with Mass spectroscopy detector as it produces easily interpreted mass spectra. A typical (shown as equation 19) reaction involving MTBSTFA as the derivatization reagent is shown below.

$$H_3C-C(CH_3)_2-Si(CH_3)_2-N(CH_3)-C(=O)-CF_3 + H-Y-R \longrightarrow H_3C-C(CH_3)_2-Si(CH_3)_2-Y-R + H_3C-N(H)-C(=O)-CF_3$$

Equation 19: Silylation reaction using N-methyl-N-t-butyldimethylsilyltrifluoroacetamide (MTBSTFA) reagent: Y = O, S, NH, NR`, COO, R, R` = Alk, Ar.

Silylation derivatization procedure using MTBSTFA can be carried out in the laboratory as follows. To a 1 - 5 mg sample, 100 µL each of reagent and solvent is added and shaken on a vortex mixture for five minutes. In case of hindered compounds, they should be heated in a closed vial at 60 °C for 1 hour. For alcohols and amines, MTBSTFA with 1 % tert-butyl dimethylchlorosilane (t-BDMCS) catalyst is used.

In another method where dicarboxylic acids are produced, 150 µL of sample, 100 µL of acetonitrile and 10 uL of MTBSTFA are added. The vial is sealed and the mixture is allowed to stand overnight. Then 100 µL of water is added to hydrolyze any unreacted MTBSTFA, followed with 250 µL of hexane. Vortex mixing and centrifuging follows before the upper hexane layer is decanted and dried to approximately 5 µL under a gentle stream of nitrogen. The sample is finally injected onto GC column for analysis.

2.3 Acylation

Derivatization by acylation is a type of reaction in which an acyl group is introduced to an organic compound. In the case of a carboxylic acid, the reaction involves the introduction of the acyl group and the loss of the hydroxyl group. Compounds that contain active hydrogens (e.g., -OH, -SH and -NH) can be converted into esters, thioesters and amides, respectively, through acylation (Zenkevich, 2009). Acylation is also a popular reaction for the production of volatile derivatives of highly polar and in volatile organic materials (Zaikin and Halket, 2003). Acylation also improves the stability of those compounds that are thermally labile by inserting protecting groups into the molecule. Acylation can render extremely polar materials such as sugars amenable to separation by GC and, consequently, are a useful alternative or complimentary to the silylation. Equation 20 shows an example of an acylation is the reaction between acetic anhydride and an alcohol.

$$CH_3OCOCOCH_3 + HOR \xrightarrow{H^+ \text{(Catalyst)}} CH_3OCOR' + HOCOCH_3$$

Equation 20: Reaction between acetic anhydride and an alcohol to produce acetate ester and acetic acid. (Blau and Halket, 1993).

Acylation has the following benefits in GC analysis.

i. It improves analyte stability by protecting unstable groups.
ii. It can provide volatility on substances such as carbohydrates or amino acids, which have many polar groups that they are nonvolatile and normally decompose on heating.

iii. It assists in chromatographic separations which might not be possible with compounds that are not suitable for GC analysis.
iv. Compounds are detectable at very low levels with an electron capture detector (ECD).

In addition, acyl derivatives tend to produce fragmentation patterns of compounds in MS applications which are clear to interpret and provide useful information on the structure of these materials. Acylation can be used as a first step to activate carboxylic acids prior to esterfication. However, acylation derivatives can be difficult to prepare especially because of interference of other reaction products (acid by-products) which need to be removed before GC separation. Acylation reagents are moisture sensitive, hazardous and odorous.

2.3.1 Derivatization reagents used in acylation

Common reagents for the Alkylation process are Fluoracylimidazoles, Fluorinated Anhydrides, N-Methyl-bis(trifluoroacetamide) (MBTFA), Pentafluorobenzoyl Chloride (PFBCI) and Pentafluoropropanol (PFPOH). Acylating reagents readily target highly polar, multi-functional compounds, such as carbohydrates and amino acids. In addition, acylating reagents offer the distinct advantage of introducing electron-capturing groups (Kataoka, 2005), and therefore enhancing detectability during analysis. These reagents are available as acid anhydrides, acyl derivatives, or acyl halides. The acyl halides and acyl derivatives are highly reactive and may be suited for use where issues of steric hindrance may be a factor. Acid anhydrides are available in a number of fluorinated configurations, which can improve detection. These fluorinated anhydride derivatives are used primarily for electron capture detection (ECD), but can also be used for flame ionization detection (FID). Fluorinated anhydrides are often used in derivatizing samples to confirm drugs of abuse. Despite this special use, their acidic nature requires that any excess or byproducts be removed prior to GC analysis to prevent deterioration of the column. Because of the acidic byproducts, the derivatization process has been carried out in pyridine, tetrahydrofuran or another solvent capable of accepting the acid by-product.

2.3.1.1 Fluorinated anhydrides

Fluorinated Anhydrides include the following compounds; Trifluoroacetoic Anhydride (TFAA), Pentafluoropropionic Anhydride (PFPA) and Heptafluorobutyric Anhydride (HFBA) which are suitable for both Flame ionization Detectors (FID) and Electron Capture Detectors (ECD). Equations i, ii, and iii representing acylation reactions using HFBA, PFPA and TFAA derivatization reagents respectively are shown below (Equation 21).

i. $C_3F_7OCOCOC_3F_7 + H\text{-}Y\text{-}R \rightarrow C_3F_7OC\text{-}Y\text{-}R + C_3F_7OC\text{-}OH$

ii. $C_2F_5OCOCOC_2F_5 + H\text{-}Y\text{-}R \rightarrow C_2F_5OC\text{-}Y\text{-}R + C_2F_5OC\text{-}OH$

iii. $CF_3OCOCOCF_3 + H\text{-}Y\text{-}R \rightarrow CF_3OC\text{-}Y\text{-}R + CF_3OC\text{-}OH$

Equation 21: Derivatization reactions using HFBA, PFPA and TFAA respectively: Y = O, NH, NR′, R, R′ = Alk, Ar.

The perfluoroacid anhydrides and acyl halide reagents react to form acidic byproducts which must be removed prior to the GC analysis in order to prevent damage to the chromatography column. The reagents basically react with alcohols, amines, and phenols to produce stable and highly volatile derivatives. Bases, such as triethylamine, can be added as an acid receptor and promote reactivity (Lin et al., 2008). Amine bases also may be used as catalysts/acid acceptors.

The following procedure by Palmer et al., (2000) for derivatization using Fluorinated Anhydrides involves the use of Triethylamine (TMA), as a catalyst. In a 5 mL vial, 50 µg (or 250 µg in case of FID) of sample is dissolve in 0.5 mL benzene. Then, 0.1 mL of 0.05 M triethylamine in benzene is added followed by 10 µL of desired anhydride such as heptafluorobutyric Anhydride (HFBA). The vial is capped and heat to 50 °C for 15 minutes. This is followed by cooling and addition of 1 mL of a 5 % aqueous ammonia solution. The cool mixture is shaken for 5 minutes, to separate the benzene layer, and injected directly onto the GC column.

2.3.1.2 Fluoracylimidazoles

Fluoracylimidazoles include Trifluoroacetylimidazole (TFAI), Pentafluoropropanylimidazole (PFPI) and Heptafluorobutyrylimidazole (HFBI) which reacts under mild conditions. The by-products (imidazole and/or N-methytrifluoroacetamide) are not acidic and therefore do not harm the column. Care must be taken because these reagents react violently with water. Fluoracylimidazoles work best with amines and hydroxy compounds (Kataoka, 2005). For example Heptafluorobutyrylimidazole readily forms derivatives with phenols, alcohols and amines (as shown in equation 22) and the derivatives are suitable for ECD.

$$\text{(imidazole)NOCC}_3\text{F}_7 + \text{H-Y-R} \longrightarrow \text{C}_3\text{F}_7\text{OC-Y-R} + \text{(imidazole)NH}$$

Equation 22: Derivatization reaction using Heptafluorobutyrylimidazole: Y = O, NH, NR´, R, R´ = Alk, Ar.

The following procedure can be used for the derivatization of alcohols, amines, amides, and phenols with imidazoles. To 1- 2 mg of sample in a vial, 2 mL toluene is added followed by 0.2 mL of imidazole (HFBI). The vial is capped and heat to 60 °C for 20 minutes. This is followed by cooling the mixture and then washing it 3 times with 2 mL of H_2O. The mixture is then dried over $MgSO_4$ and inject onto the GC for separation.

2.3.1.3 N-Methyl-bis(trifluoroacetamide) (MBTFA)

N-Methyl-bis(trifluoroacetamide) (MBTFA) reagent reacts rapidly with primary and secondary amines, and also slowly with hydroxyl groups and thiols. Reaction conditions are mild with relatively inert and non acidic by-products and therefore do not damage the GC column (Lelowx et al., 1989). The general reaction is presented in equation 23:

$$F_3COCN\ (CH_3)\ OCCF_3 + H\text{-}Y\text{-}R \rightarrow F_3OC\text{-}Y\text{-}R + CH_3NHOCCF_3$$

Equation 23: Representative reaction of the derivatization of amines, hydroxyl groups and thiols using N-Methyl-bis(trifluoroacetamide) (MBTFA) reagent: Y = O, S, NH, NR´, R, R´ = Alk, Ar.

N-Methyl-N-bis(trifluoroacetamide is recommended for the analysis of sugars (Matsuhisa, 2000; Hannestad, 1997) and as an acylation reaction is often used for amine drugs, such as stimulants, amino acids, and alcohols.

In the method used by Matsuhisa, (2000), 20 µL of D-glucose standard aqueous solution (10 µg/mL) was sampled and freeze-dried. Trifluoroacetylation was performed by adding 10 µL of pyridine and 10 µL of MBTFA (N-methyl-bis-trifluoroacetamide) to the dried sample, and then heating at 60 °C for 60 minutes. The cooled sample was ready for GC analysis.

Donike, (1973) applied the following procedure for the rapid derivatization of amines. To 10 mg sample, 0.2 mL MBTFA and 0.5 mL solvent of choice is added in a reaction vial. The vial is capped and heat to 60 - 100 °C for 15 to 30 minutes. It was also observed that some compounds with steric hindrance require additional heating.

2.3.1.4 Pentafluorobenzoyl Chloride (PFBCI)

Pentafluorobenzoyl chloride (PFBCI) is used in making derivatives of alcohols and secondary amines of which secondary amines are the most highly reactive, forming the most sensitive ECD derivatives of amine and phenol. Phenols are the most receptive site for this reagent (Regis, 1999). Pentafluorobenzoyl chloride (PFBCI) is suitable for functional groups that are sterically hindered. A base such as NaOH is often used to remove the HCl that is produced as byproduct. This derivatization procedure which is presented by the reaction below (equation 24) basically uses a pentafluorobenzoyl chloride (PFBCl) to provide rapid formation of the derivatives of amines and phenols.

$$C_6\,F_5\text{-OCCl} + C_6H_5\text{-OH} \rightarrow C_6\,F_5\text{-OCO-}\,C_6H_5 + HCl$$

Equation 24: Formation of derivatives of amines and phenols using Pentafluorobenzoyl chloride (PFBCI).

The following derivatization procedure can be used for PFBCI reagent. In a 5 mL vial, 25 - 50 mg of sample and 2.5 mL of 2.5 N NaOH are dissolved. Then, 0.1 g of PFBCl, is added and the vial is capped, shaken vigorously for 5 minutes. The derivative formed is extracted into methyl chloride and dried over $MgSO_4$. The sample is then injected directly onto the GC column.

2.3.1.5 Pentafluoropropanol (PFPOH)

Pentafluoropropanol ($CF_3CF_2CH_2OH$) is used in combination with pentafluoropropionic anhydride (PFPA) and is applied commonly with polyfunctional bio-organic compounds (Regis, 1999). For example, 2,2,3,3,3-Pentafluoropropanol is used in combination with PFPA to make derivatives of the most common functional groups, especially polyfunctional bio-organic compounds (Regis, 1999). The formed derivatives are highly suitable for ECD.

Sample analysis and derivatization can be done through the following sequential derivatization procedure, in which 1 mg of sample is added in a reaction vial, followed by the addition of 50 µL PFPOH and 200 µL of PFPA. The sample is then heated to 75 °C for 15 minutes and then evaporated to dryness under N_2. An additional 100 µL PFPA is added and the sample is again heated to 75 °C for 5 minutes. The sample is finally evaporated to dryness and dissolved in ethyl acetate prior to injection.

2.3.1.6 4-Carbethoxyhexafluorobutyryl chloride (4-CB)

4-Carbethoxyhexafluorobutyryl chloride (4-CB) forms stable products with secondary amines, such as methamphetamine, allowing the removal of excess agent by adding protic solvents. 4-Carbethoxyhexafluorobutyryl chloride decreases the net charge of the peptides and increased hydrophobicity (Klette, 2005). The following equation 25 shows the derivatization of an amine using 4-CB reagent.

$$ClOCC_3F_6OCOC_2H_5 + R\text{-}NH_2 \rightarrow RNHOCC_3F_6OCOC_2H_5 + HCL$$

Equation 25: Amine derivatization using 4-Carbethoxyhexafluorobutyryl chloride (4-CB) reagent.

The procedure for derivatization follows the one used by Dasgupta et al., (1997). In the procedure, 50 µL of 4-CB is added to dried sample (dried by passing of a gentle stream of nitrogen gas to remove the organic phase) followed by incubation at 80 °C for 20 minutes. After derivatization the excess 4-CB is evaporated and the derivatized sample is reconstituted in 50 µL ethyl acetate of which 1 - 2 µL is injected to GC for analysis.

3. GC Chiral Derivatization

Chiral derivatization involves reaction of an enatiomeric molecule with an enatiomerically pure chiral derivatizing agent (CDA) to form two diastereomeric derivatives that can be separated in this case using GC. A solution in which both enantiomers of a compound are present in equal amounts is called a racemic mixture. Diastereomers are stereoisomers (they have two or more stereo centers) that are not related as object and mirror image and are therefore not enantiomers. In other word, unlike enatiomers which are mirror images of each other and non-sumperimposable, diastereomers are not mirror images of each other and non-superimposable. Diastereomers can have different physical properties and reactivity.

Any molecule having asymmetric carbon is called as chiral molecule. Chirality of analyte molecules requires special consideration in their analysis and separation techniques. Scientists and other regulatory authorities are in the demand of data on concentrations and toxicity of the chiral pollutants, and therefore chiral derivatization is becoming an essential, urgent and demanding field. However the derivatization procedures are tedious and time consuming due to the different reaction rates of the individual enantiomers (Schurig, 2001).

Generally, there are two ways of separating enantiomers by chromatography:

i. Separation on an optically active stationary phase
ii. Preparation of diastereomeric derivatives that can be separated on a non chiral stationary phase.

The second option (ii above) requires derivatization of the analyte molecule. The presence of a suitable functional group in a pollutant is a condition for a successful derivatization of a chiral molecule. The indirect chromatographic analysis of racemic mixtures can be achieved by derivatization with a chiral derivatizing agent resulting into the formation of diastereoisomeric complex/salt (Hassan et al., 2004). From the resulting chromatograms, calculations are made to determine the enantiomeric concentration of the analyte. The diastereoisomers having different physical and chemical properties can be separated from each other by an achiral chromatographic method. Nowadays, the chromatographic methods are the most popular for enantiomeric analysis of environmental pollutants (Hassan et al 2004).

3.1 Gas chromatography chiral derivatization reagents

Gas Chromatography analysis of enantiomeric compounds on nonracemic or achiral stationary phases requires the use of enantiopure derivatization reagents (Hassan et al, 2004). Enantiopure compounds refer to samples that contain molecules having one chirality within the limits of detection. These reagents generally target one specific functional group to produce diastereomers of each of the enantiomeric analytes in GC to produce chromatograms. Some of the most common Gass Chromatography chiral derivatization reagents are: (-) menthylchloroformate (MCF), (S)-(−)-N-(Trifluoroacetyl)-prolylchlorides (TPC), (−)-α-Methoxy-α-trifluoromethylphenylacetic acid (MTPA). Examples of the use of these derivatizing agents were employed in analysis of major drugs of forensic interest, and for optically active alcohols (Tagliaro et al., (1998).

3.1.1 N-trifluoroacetyl-L-prolyl chloride (TPC)

The reagent is used for optically active amines, most notably amphetamines as represented in the following reaction (equation 26) where N-Trifluoroacetyl-L-prolyl chloride couples with amines to form diastereomers which can be separated on GC columns as it increases the sample volatility.

Equation 26: N-Trifluoroacetyl-L-prolyl chloride derivatization of amines

The standard TPC derivatization procedure is as follows; following sample clean up and extraction procedure, 1.0 mL of the TPC reagent is added to the clean sample. The mixture is allowed to stand for 5 min before the addition of 20 μL of triethylamine to take up excess unreacted TPC. After 15 min of intermittent shaking, 1.0 mL of 6 N HC1 is used to remove the ammonium salt. The mixture is finally washed with 1 mL of distilled water and then dried over anhydrous magnesium sulfate before dilution and analyzed by GC.

3.1.2 (S)-(−)-N-(Trifluoroacetyl)-prolylchloride (l-TPC)

(S)- (−)-N-(Trifluoroacetyl)-prolylchloride (l-TPC) is widely used for amine drugs analysis. Qiao Feng Tao & Su Zeng (2002) analyzed the enantiomers of chiral amine drugs by using stereo selective methods to separate enantiomers on an achiral capillary gas chromatography by pre-column chiral derivatization with S-(_)-N-(fluoroacyl)-prolyl chloride. In the study, it was noted that the stereo selectivity and sensitivity can be improved by chiral derivatization. The method has been used to determine S-(+) - methamphetamine in human forensic samples and to analyze enantiomers of amphetamine and fenfluramine in rat liver microsomes. Two analytical procedures used by Qiao Fang Tao and Su Zing (2002) for aqueous and organic phase derivatization methods are described below.

Aqueous phase derivatization method: Urine or liver homogenate was added to a 10 ml Teflon-lined, screw-capped test tube. The pH of the sample was adjusted to 9 by 10 M NaOH, then 0.5 ml of chilled 1 M sodium bicarbonate/sodium carbonate buffer (pH 9.0). After mixing, 40 µL of S-(_)-TFAPC was added, and the tube was vortex mixed for 30 min at room temperature. The resulting mixture was saturated with NaCl and adjusted to pH 9, followed by addition of 0.5 ml of chilled 1 M sodium bicarbonate/sodium carbonate buffer (pH 9.0) and extracted with 4 ml of ethyl acetate by gently shaking for 15 min. After the phases separate, the organic layer was washed with 2 ml of deionized water. The organic layer was evaporated to dryness under a gentle stream of air at 40 °C. The residue was cooled to room temperature and reconstituted with ethyl acetate. An aliquot of 1 µL of analyte was analyzed by GC/MS.

Organic phase derivatization method: A 1.0 ml sample of microsomal mixture spiked with chiral amine was piped into a 15 ml screw capped test tube. The pH of the sample was adjusted to 12 - 13 with 40% NaOH. The mixtures were extracted with 2 ml of chloroform by rotatory shaking for 1 min. After centrifugation at 4000 rpm for 10 min, the aqueous phase was removed by aspiration. The remaining organic phase was dried by anhydrous sodium sulfate, and then transferred to another clean screw-capped test tube. A 10 µL sample of triethylamine as a catalytic agent and 40 µL S-(_)-TFAPC were added into the test tube and mixed. The tube was capped and allowed to react for 15 min at room temperature by gentle shaking. The chloroform layer was washed with 2 ml distilled water and evaporated to dryness at under a water bath and a gentle stream of air 50° C. The residues were allowed to cool to room temperature and reconstituted with 40 µL ethyl acetate prior to injection into the GC/FID system (Qiao Feng Tao & Su Zeng, 2002).

3.1.3 (–)-α-Methoxy-α-trifluoromethylphenylacetic acid (MTPA)

(-)-α-Methoxy-α-trifluoromethylphenylacetic acid is also mostly used in drug analysis. In a study by S-M Wang et al., (2005) pharmaceutical drugs were analyzed using compounds that were derivatized with (-)-α-Methoxy-α-trifluoromethylphenylacetic acid (MTPA) as the derivatizing reagent. These types of derivatization reactions are represented by equation 27. In the reaction the hydroxyl group on the MTPA molecule is lost in the derivative formation.

$$\mathrm{HO{-}\overset{\overset{\large O}{\|}}{C}{-}\underset{\underset{\large CF_3}{|}}{\overset{\overset{\large C_6H_5}{|}}{C}}{-}O{-}CH_3 + R{-}NH_2 \longrightarrow RHN{-}\overset{\overset{\large O}{\|}}{C}{-}\underset{\underset{\large CF_3}{|}}{\overset{\overset{\large C_6H_5}{|}}{C}}{-}O{-}CH_3 + H_2O}$$

Equation 27: Derivatization reaction between (-)-α-Methoxy-α-trifluoromethylphenylacetic acid and an amine group of an analyte molecule.

The derivatization using MTPA can be achieved as follows: after extraction, clean up and drying the sample under nitrogen, the residue is added 50 µL N, N-dicyclohexycarbodiimide and 100 µL MTPA. The reaction mixture is then thoroughly mixed, then incubated at 70 °C for 20 min before GC analysis.

4. Summary

The choice of the derivatization technique for analysis of compounds will depend on the available reagent and reaction types that can produce derivatives that give desirable results in GC. The derivatives must be suitable, detectable and efficient for GC analysis. The evaluation of the functional group of the analyte, the GC detector and even the by products of the reaction among others considerations will guide the choice of derivatization technique.

For example, insertion of perfluoroacyl groups into a molecule enhances its detectability by electron capture. The presence of a carbonyl group adjacent to halogenated carbons in an analyte enhances the electron capture detector (ECD) response. For mass spectroscopic detector, acyl derivatives tend to direct the fragmentation patterns of compounds and therefore, providing useful information on the structure of molecules.

For acid analytes, the first choice for derivatization is esterification. Acids are reactive compounds and also more polar to be separated well by gas chromatography. The underivatized acids will tend to tail because of the adsorption and non-specific interaction with the column. Esterification is used to derivatize carboxylic acids and other acidic functional groups. The reaction involves the condensation of the carboxyl group of an acid and the hydroxyl group of an alcohol, with the elimination of water. Consequently, the polarity of the molecule is reduced.

Nearly all functional groups which present a problem in gas chromatographic separation (hydroxyl, carboxylic acid, amine, thiol, phosphate) can be derivatized by silylation reagents. The derivatives of the silylation reactions are generally less polar, more volatile and more thermally stable. The introduction of a silyl group(s) can also serve to enhance mass spectrometric properties of derivatives, by producing either more favorable diagnostic fragmentation patterns of use in structure investigations, or characteristic ions of use in trace analyses in other related techniques.

In considering the functional group or compound type, dimethylformamide-dialkylacetals (DMF-DEA) or similar type of reagents are recommended for sterically hindered aldehydes, amines, carboxylic acids, and phenols. Shorter chain reagents like DMF- DEA will produce more volatile derivatives than longer chain reagents. The by products of the reaction will also determine the choice of reagent. In this view, perfluoroacylimidazole is a better choice rather than perfluoroacid anhydrides. Perfluoroacid anhydrides produce acidic byproducts which must be removed from the reaction mixture before the derivatives are injected onto the GC column. With perfluoroacylimidazole there are no acid byproducts.

The complexity surrounding chiral derivatization is generally because of different reaction rates of the individual enantiomers. Derivatization with a chiral derivatizing agent resulting into the formation of diastereoisomeric complexes is an option in GC analysis, and has been applied to analyze various analytes (Hassan et al., 2004 & Schurig, 2001). Table 1 gives a summary guide of derivatization technique and reagent selection based on the functional group of the analyte molecule.

Functional group	Reaction type	Derivatization reagent
Alcohols and Phenols	Silylation	Bis(trimethylsilyl)–acetamide, Bistrimethylsilyltrifluoroacetamide, N-methyl-N-t-butyldimethylsilyltrifluoroacetamide
	Acylation	Heptafluorobutyrylimidazole, Pentafluoropropionic Anhydride, Trifluoroacetic anhydride, N-Methyl-bis(trifluoroacetamide)
	Alkylation	Dimethylformamide, Pentafluorobenzyl bromide
Carboxylic acids	Silylation	Bis(trimethylsilyl)–acetamide, Bistrimethylsilyltrifluoroacetamide, Trimethylsilylimidazole, N-methyl-N-t-butyldimethylsilyltrifluoroacetamide
	Acylation	Pentafluoropropanol / pentafluoropropionic anhydride
	Alkylation	Dimethylformamide, Tetrabutylammonium hydroxide
Active hydrogens	Silylation	Bis(trimethylsilyl)–acetamide, Bistrimethylsilyltrifluoroacetamide / Trimethylchlorosilane, Hydrox-Sil, N-methyl-trimethylsilyltrifluoroacetamide,
	Acylation	Pentafluoropropanol / pentafluoropropionic anhydride
	Alkylation	Dimethylformamide, Tetrabutylammonium hydroxide
Carbohydrates and Sugars	Silylation	Hexamethyldisilzane, Trimethylsilylimidazole
Amides	Silylation	Bis(trimethylsilyl)–acetamide, N, O-bis-(trimethylsilyl)-trifluoroacetamide
	Acylation	Heptafluorobutyrylimidazole
	Alkylation	Dimethylformamide
Amines	Silylation	Bistrimethylsilyltrifluoroacetamide, N-methyl-N-t-butyldimethylsilyltrifluoroacetamide
	Acylation	Trifluoroacetic anhydride, Pentafluorobenzoyl chloride, Heptafluorobutyrylimidazole
	Alkylation	Dimethylformamide (Diacetals)
Amino acids	Silylation	Bistrimethylsilyltrifluoroacetamide, Trimethylsilylimidazole
	Acylation	Heptafluorobutyrylimidazole
	Alkylation	Dimethylformamide, Tetrabutylammonium hydroxide
Catecholamines	Silylation	Trimethylsilylimidazole
	Acylation	Pentafluoropropionic Anhydride, Heptafluorobutyrylimidazole,
Inorganic anions	Silylation	Bistrimethylsilyltrifluoroacetamide, N-methyl-N-t-butyldimethylsilyltrifluoroacetamide
Nitrosamine	Silylation	Bistrimethylsilyltrifluoroacetamide
	Acylation	HFBA, Pentafluoropropionic anhydride, Trifluoroacetic anhydride
	Alkylation	Dimethylformamide, Pentafluorobenzyl bromide
Sulfonamides	Acylation	Trifluoroacetoic & Heptafluorobutyric Anhydride, Pentafluorobenzyl bromide
Sulfides	silylation	Trimethylsilylimidazole

Table 1. A summary guide for the derivatization techniques and reagents selection based on sample types.

5. References

Allmyr M, McLachlan MS, Sandborgh-Englund G, Adolfsson-Erici M. (2006). Determination of Triclosan as its pentafluorobenzoyl ester in human plasma and milk using electron capture negative ionization mass spectrometry. Anal Chem. 2006; 78: 6542–6.

Bao Ming-Liang, Pantani Francesco, Griffini Osvaldo, Burrini Daniela, Santianni Daniela, Barbieri Katia (1998). Determination of carbonyl compounds in water by derivatization–solid-phase micro extraction and gas chromatographic analysis. Journal of Chromatography A. 809 (1998) 75–87

Blau, K., King, G. (1997). Handbook of Derivatives for Chromatography; Heyden & Sons Ltd.; London, 1979

Butts W.C. (1972). Two-column gas chromatography of trimethylsilyl derivatives of Biochemically significant compounds.. Analytical Biochemistry. Volume 46, Issue 1, March 1972, Pages 187-199

Chen, B.-G., Wang, S.-M. and Liu, R. H. (2007). GC-MS analysis of multiply derivatized opioids in urine. Journal of mass spectrometry JMS (2007). Volume: 42, Issue: 8, Pages: 1012-1023

Chien, C.-J.; Charles, M. J.; Sexton, K. G.; Jeffries, H. E. (1998). Analysis of airborne carboxylic acids and phenols as their pentafluorobenzyl derivatives: gas Chromatography/Ion Trap Mass Spectrometry with a novel chemical ionization reagent, PFBOH, Environmental Science And Technology. 1998, 32, 299-309.

Christie, W.W. (1993): preparation of ester derivatives of fatty acids for chromatographic analysis. Advances in Lipid Methodology - *Two*, pp. 69-111 (1993) (Ed. W.W. Christie, Oily Press, Dundee),

Danielson, Neil D., Gallagher Patricia A., and Bao James J. (2000). Chemical Reagents and Derivatization Procedures in Drug Analysis. Encyclopedia of Analytical Chemistry. John Wiley & Sons Ltd, Chichester, 2000. R.A. Meyers (Ed.). pp. 7042–7076.

Dasgupta, Amitava, Blackwell Walter and Burns Elizabeth (1997). Gas Chromatographic-mass spectrometric identification and quantitization of urinary phenols after derivatization with 4-carbethoxyhexafluorobutyryl chloride, a novel derivative. Journal of Chromatography B: Biomedical Sciences and Applications Volume 689, Issue 2, 21 February 1997, Pages 415-421

Destaillats, H.; Spaulding, R. S.; Charles, M. J. (2002). Ambient air measurement of acrolein and other carbonyls at the Oakland-San Francisco Bay Bridge toll plaza. Environmental Science and Technology 2002, 36, 2227-2235.

Donike, M. (1973). Acylation with bis (acylamides). N-methyl-bus (trifluoroacetamide) and bis (trifluoroacetamide), two new reagents for trifluoroacetylation. Journal of Chromatography A. Volume 78, Issue 2, 25 April 1973, Pages 273-279.

Galceran, M. T.; Moyano, E.; Poza, J. M. (1995). Pentafluorobenzyl derivatives for the gas-chromatographic determination of hydroxy-polycyclic aromatic hydrocarbons in urban aerosols. Journal of Chromatography A 1995, 710, 139-147.

Gehrke, Charles W. (1968). Quantitative gas-liquid chromatography of amino acids in proteins and biological substances: Macro, semimicro, and micro methods. Analytical Biochemistry Laboratories (Columbia, Mo). 1st edition. 101-289-264

Hannestad, U.; Lundblad, A. (1997). Accurate and Precise Isotope Dilution Mass Spectrometry Method for Determining Glucose in Whole Blood. Clinical Chemistry, 1997, 43, 5, 794-800.

Hassan, Y. Aboul-Enein & Imran Ali (2004): Analysis of the chiral pollutants by Chromatography, Toxicological & Environmental Chemistry, 86:1, 1-22

Isoherranen, Nina and Soback Stefan (2000). Determination of gentamicin after Trimethylsilylimidazole and trifluoroacetic anhydride derivatization using gas Chromatography and negative ion chemical ionization ion trap mass spectrometry. Analyst, 2000, 125, 1573–1576

Jang Myoseon and Kamens Richards, M. (2001). Characterization of Secondary Aerosol from the Photooxidation of Toluene in the Presence of NOx and 1-Propene Environ. Sci. Technol. 2001, 35, 3626-3639

Kataoka, Hiroyuki (2005). Gas chromatography of amines as various derivatives. Quantitization of amino acids and amines by chromatography - methods and protocols. Journal of Chromatography Library. Volume 70, 2005, Pages 364-404.

Kawahara, F.K. (1968). Micro determination of pentafluorobenzyl ester derivatives of Organic acids by means of electron capture gas chromatography.

Klette, L Kevin., Jamerson H. Matthew, Morris-Kukoski L. Cynthia, Kettle R. Aaron, and Snyder J. Jacob (2005). Rapid Simultaneous Determination of Amphetamine, Methamphetamine, 3, 4-Methylenedioxyamphetamine, 3, 4-Methylenedioxymethamphetamine, and 3, 4-Methylenedioxyethyla. Journal of Analytical Toxicology. Volume Number: 29: Issue 7: 669–674

Knapp, Daniel R. (1979). Handbook of Analytical Derivatization Reaction; pages 10, 10; Wiley & Sons; New York, 1979.

Kühnel, E., Laffan Dr, D.D.P., Lloyd-Jones, G.C., Martínez del Campo, T., Shepperson, I.R. Slaughter,J.L.,(2007). Mechanism of Methyl Esterification of Carboxylic Acids by Trimethylsilyldiazomethane. Angewandte Chemie International Edition. Volume 46, Issue 37, pages 7075–7078, September 17, 2007.

Kuo, H.W., Ding, W.H. (2004). Trace determination of bisphenol A and phytoestrogens in infant formula powders by gas chromatography-mass spectrometry. J. Chromatogr. A, 1027, 67.

Lelowx, M.S. De jong E.G and Meas RA.A. (1989). Improved screening method for beta-blockers in urine using solid- phase extraction and capillary gas chromatography—mass spectrometry. Journal of Chromatography. 488 (1989). 357-.367.

Lin, D.-L. Wang, S.-M. Wu, C.-H. Chen, B.-G. Liu, R.H. (2008). Chemical Derivatization for the Analysis of Drugs by GC- MS — A Conceptual Review. Journal of Food and Drug Analysis, Vol. 16, No. 1, 2008, Pages 1-1

Matsuhisa, M.; Yamasaki, Y.; Shiba, Y.; Nakahara, I.; Kuroda, A.; Tomita, T. (2000). Important Role of the Hepatic Vagus Nerve in Glucose Uptake and Production by the Liver Metabolism, 2000, 49, No 1 (January), 11-16.

Naritisin, D.B. and Markey, S.P. (1996). Assessment of DNA oxidative damage by Quantification of thymidine glycol residues using gas chromatography/electron Capture negative ionization mass spectrometry. Anal. Biochem. 241:35-41.

Orata, F. Quinete, N. Wilken, R.D. (2009). Long Chain Perfluorinated Alkyl Acids Derivatization and Identification in Biota and Abiota Matrices Using Gas

Chromatography. Bulletin of Environmental Contamination and Toxicology 2009 Vol. 83 No. 5 pp. 630-635

Palmer, R.B, Kim N.H. and Dasgupta A (2000). Simultaneous determination of fenfluramine and phentermine in urine using gas chromatography-mass spectrometry with pentafluoropropionic anhydride derivatisation. Ther. Drug Monit. 22 (2000). 418-422.

Pierce, Alan E. (1968). Silylation of Organic Compounds. Pierce Chemical Company, 1968. Pierce, (2004) .GC Derivatization. .Applications Handbook & Catalog.

Qiao Feng Tao and Su Zeng (2002). Analysis of enantiomers of chiral phenethylamine drugs by capillary gas chromatography/mass spectrometry/flame-ionization detection and pre-column chiral derivatization. J. Biochem. Biophys. Methods 54 (2002) 103–113.

Rao, X.; Kobayashi, R.; White-Morris, R.; Spaulding, R.; Frazey, P.; Charles, M. J. (2001). J GC/ITMS measurement of carbonyls and multifunctional carbonyls in PM2.5 particles emitted from motor vehicles.Journal of AOAC International 2001, 84, 699-705.

Ribeiro Bárbara, Guedes de Pinho Paula, Andrade B. Paula, Baptista Paula, Valentão Patrícia. (2009). Fatty acid composition of wild edible mushrooms species: A comparative study. Micro chemical Journal 93 (2009) 29–35 Regis (1999). Chromatography Catalog, pages 89, 90

Saraji, M. and Mirmahdieh, S. (2009). Single-drop micro extraction Followed by in-syringe derivatization and GC-MS detection for the determination of parabens in water and cosmetic products. J. Sep. Sci. 2009, 32, 988 - 995

Schauer, J. J.; Kleeman, M. J.; Cass, G. R.; Simoneit, B. R. T. (2002). Environ. Science and Technology 2002, 36, 1169-1180.

Schurig V (2001). Chiral separation of amino acids is of great importance in biology, Pharmaceutics and agriculture. This has been performed traditionally with Chromatographic methods such as gas chromatography (GC). J. Chromatogr. A 906 (2001) 275.

Scott, R. P.W. (2003). Gas chromatography. book 2 Chrom-Ed Book Series. http://www.library4science.com/eula.html. 2

S-M Wang, Lewis j. Russell, Canfield Dennis, Li Tien-Lai, Chen Chang-Yu, Liu H. Ray (2005). Enantiomeric determination of ephedrines and norephedrines by chiral derivatization gas chromatography–mass spectrometry approaches. *J. Chromatogr. B 825 (2005) 88–95*

Sobolevsky, T.G. Alexander I.R. Miller, B. Oriedo, V., Chernetsova, E.S., Revelsky, I. A. (2003). Comparison of silylation and esterification/acylation procedures in GC-MS analysis of amino acids. Journal of Separation Science. Volume 26, Issue 17, pages 1474–1478, November 2003

Spaulding, R. S.; Schade, G. W.; Goldstein, A. H.; Charles, M. J. Journal of Geophysical Research, [Atmospheres] 2003, 108, ACH7/1-ACH7/17.

Szyrwińska, K. Kołodziejczak A., Rykowska I., Wasiak W., and Lulek J. (2007). Derivatization and Gas chromatography- low-resolution Mass Spectrometry of Bisphenol A. Acta Chromatographica, No. 18, 2007

Tagliaro Franco, Smith Frederick P, Tedeschi Luciano, Castagna Franca, Dobosz Marina, Boschi Ilaria and Pascali Vincenzo (1998). Toxicological and Forensic Applications Chapter 21. Journal of Chromatography Library Volume 60, 1998, Pages 917-961

Thenot, J.-P., Horning, E.C. (1972). Amino acid *N*-dimethyllaminomethylene alkyl esters. New derivatives for GC and GC-MS studies. Anal. Lett.5 (8): 519-29.

Thenot, J. P, Stafford M. and Horning M. G. (1972). Fatty acid esterification with N, N-Dimethylformamide dialkyl acetals for GC analysis. Anal. Lett. 5(4): 217-23.

Zaikin, V.G., Halket, J.M (2003). Derivatization in mass spectrometry--2. Acylation. Eur J Mass Spectrum. 2003; 9(5):421-34.

Zenkevich, Igor, G. (2009). Acids: Derivatization for GC Analysis Encyclopedia of Chromatography, Third Edition 2009, DOI: 10.1081/E-ECHR3-120045222

Porous Polymer Monolith Microextraction Platform for Online GC-MS Applications

Samuel M. Mugo, Lauren Huybregts, Ting Zhou and Karl Ayton
Grant MacEwan University, Edmonton, Canada

1. Introduction

One of the key steps in chemical analysis is sample preparation. It is a very important step prior to analysis, otherwise the analyte signal could be suppressed by sample matrix interference. In most cases sample preparation entails some form of extraction of the analyte(s) of interest from the interfering species. Conventional extraction techniques include liquid-liquid extraction (LLE) and solid phase extraction (SPE). LLE involves mixing two immiscible solvent together, and based on solubility equilibria the analyte or the interfering species partitions preferentially in the organic solvent. Ideally, the partition coefficient is very high to allow for efficient extraction (Park et al., 2001). In SPE, analytes of interest are extracted from solution by passing the liquid through a solid phase, such as a C-18 SPE cartridge. Based on different physical and chemical properties, the desired solute either is adsorbed onto the solid phase, from whence it can later be eluted, or remains in solution while impurities are retained on the solid phase (Körner et al., 2000). The LLE and SPE extraction methods are laborious and demand the use of copious amounts of samples and solvents, therefore driving up the cost of chemical analysis. For LLE, emulsion formation can be a nuisance. These methods are also limited by factors like cartridge clogging in SPE, single use of the SPE and the need for toxic and polluting solvents (such as halogenated solvents like chloroform and dichloromethane) in LLE.

Other emerging extraction techniques that are based on miniaturization of LLE in effort to minimize solvent use and simplify the sample preparation process include: single-drop microextraction (SDME), hollow fibre liquid-phase microextraction (HF-LPME) and dispersive liquid–liquid microextraction (DLLME). Briefly, SDME, introduced by Cantwell and Jeannot in 1996, entails exposing a microdrop of extractant placed on the end of a Teflon rod to the sample solution or sample headspace. The HF-LPME uses two phase sampling mode, a hydrophobic hollow fibre is used to support the extractant and the fibre is exposed to the sample and the analyte is extracted in the organic extractant. On the other hand, the DLLME is based on rapidly injecting a few microliters of high density solvents mixed with a disperser solvent into the aqueous sample. The turbulence from the rapid injection causes large interfacial area emulsion droplets to form and distribute in the sample solution. Analytes partition into this emulsion droplets; and after centrifugation, they sediment at the bottom of a vial and then are sent for analysis. In general these miniaturized LLE techniques are convenient, efficient (given a high partition coefficient), inexpensive and requires very little solvent. They are also easily coupled to gas chromatography (GC) (Jeannot & Cantwell, 1996). An extensive discussion of these methods is available and covered in the review by Pena-Pereira et al., 2009. An obvious disadvantage with these methods is that only a few organic solvents are usable; and so the methods lack versatility and are only applicable to the extraction of a limited number of analytes.

Solid phase microextraction (SPME), developed by Pawliszyn in the 1990s (Arthur & Pawliszyn, 1990), surmounts some of the limitations of the SPE, LLE and miniaturized LLE. SPME method employs a fused silica fiber coated with a polymeric coating to extract organic compounds from their sample matrix and directly transfer analytes into GC, eliminating offline sample extraction steps (Z. Zhang & Pawliszyn, 1995). Different polymer stationary phases have been used in SPME for the analysis of many different analyte classes. Common commercially-available SPME stationary phases include polyacrylate, polydimethylsiloxane (PDMS), divinylbenzene (DVB)/PDMS, carboxen/PDMS, and DVB/Carboxen/PDMS. However, SPME fibres are generally very fragile and cases of them breaking off in the GC port are very common. Therefore they have to be used with extreme care, and sometimes they have to be replaced fairly often, which makes them expensive. Another disadvantage of SPME is the limited variety of available stationary phases, which limits the range of analyte classes that can be extracted. There have been publications of novel SPME stationary phases being developed to address this.

Second and third generation variants of SPME have been developed, such as the needle-trap devices. The need-trap extraction method involves packing a needle with polymer beads or carbon based sorbents. A gaseous sample is drawn through the needle housing the sorbent plug and the analyte of interest is adsorbed and then directly injected and desorbed into GC (Ueta et al., 2010). The needle-trap extraction method possesses the benefits of microscale extraction (*i.e.* low cost, high efficiency, low-volume solvent use) while overcoming SPME's problem of fragility issues. However, this method is only applicable to headspace extraction of volatile samples and so can only be used to extract a limited range of analytes.

A new class of microextraction technique is polymer monolith microextraction (PMME) introduced by Feng and coworkers in 2006. Feng and coworkers fabricated their device using a regular plastic syringe (1 ml) and a poly (methacrylic acid-co-ethylene glycol dimethacrylate) monolith entrained in a silica capillary (2 cm × 530 µm I.D.) (Zhang et al. 2006). PPME shares the advantages of SPME while solving most of its limitations. In general the most important component of the PPME device is the porous polymer monolith usually formed in the confines of a silica capillary, although other materials such as polyether ether ketone (PEEK) and titanium tubes can used (Wen et al. 2006).

Porous polymer monoliths are relatively new materials synthesized by carrying out free radical (radicals generated by heat or ultra-violet radiation following destabilization of azo initiators) polymerization of cross-linking and monovinyl monomers, in the presence of suitable porogenic solvents. The generated radicals initiate a rapid polymer chain growth at what become nucleation sites, which continue to grow as the reaction proceeds. As polymer molecular weight increases, the solubility decreases and a two-phase system of solid polymer and liquid solvent results. The resulting monolith microstructure consists of an agglomeration or globules, whose size directly impacts resulting pore size distribution.

Globule size is influenced by many factors, including the number of nucleation sites present, monomer concentration, solubility, and degree of cross-linking. Thus, polymer microstructure can be controlled by rate of reaction, monomer/porogen ratio, type of porogenic solvents and fraction of cross-linking monomers. Detailed descriptions of various chemistries, and characterization of resulting monoliths can be found in references (Gibson et al. 2008 & Svec, 2010).

The polymer monoliths used in making the PPME platforms can easily be made from different types of monomers with different polarities depending on the analytes of interests. The large surface area provided by the tailor-able macropores in monolith structure may help to improve the extraction efficiency. The convection flow provided by the flow-through channel within the monolith also helps in accelerating mass transfer. Furthermore, PPME device is very easy to prepare, which also makes it a better choice than particle packed extraction cartridges.

There have been numerous literatures reporting on the application of this PMME method for extraction and coupling to GC and LC, with the latter being more common. For example, Wen et.al detected and quantified a series of sexual hormones (testosterone, methyltestosterone and progesterone) in liquid cosmetics extracted by a poly (methacrylic acid-co-ethylene glycol dimethacrylate) monolithic capillary. By simple dilution and filtering after extraction, the samples were directly injected into HPLC system. The limit of detection was 2.3 to 4.6 μg/L. Calibration curves showed good linearity in the concentration range of 10 to 1000 μg/L with R^2 above 0.996 (Wen et al., 2006).

A variation on conventional PPME involves the production of molecularly-imprinted polymers (MIPs) to make up the extraction monolith (Bravo et al., 2005; Sanbe & Haginaka, 2003; Schweitz, 2002; Zhou et al., 2010). The imprints left by template molecules act as "artificial receptor-like binding sites" which are selective to the analyte of interest. Zhou and coworkers synthesized a molecularly imprinted solid-phase microextraction monolith (MIP-PPME) for selective extraction of pirimicarb in tomato and pear. They prepared a pirimicarb MIP monolith in a micropipette tip using methacrylic acid as the functional monomer, ethylene dimethacrylate as the cross-linker and the mixture of toluene-dodecanol as the porogenic solvent. The dynamic linear ranges were from 2.0 to 1400 μg/kg for pirimicarb in tomato and pear (Zhou et al., 2010).

Until now polymer monolithic materials used in PMME have mainly been poly-(methacrylic acid-co-ethylene dimethacrylate) (MAA-co-EDMA) (Liu et al., 2011) and most documented applications have been for LC. Our group has demonstrated the use of many other types of acrylate, epoxy-based, and acrylamide polymer monoliths for PMME and have shown that these simple PPME devices can be fabricated for online-GC applications. A schematic set up on how they are coupled to GC-MS is shown in Figure 1, which is similar to a commercial SPME operation. The fabrication and application of these newly developed PMME materials will be discussed in this chapter. In particular we shall demonstrate the application of PPME to the extraction of caffeine, polyaromatic hydrocarbons (PAHs) and hormones in water. We have also evaluated the use of porous layer open tubular (PLOT) monoliths as PPME platforms. The potential advantage of PLOT (otherwise will be referred in this chapter as open tubular polymer monolith) is the faster mass transfer and consequently faster extraction than PPME. These materials are still being evaluated in our research group.

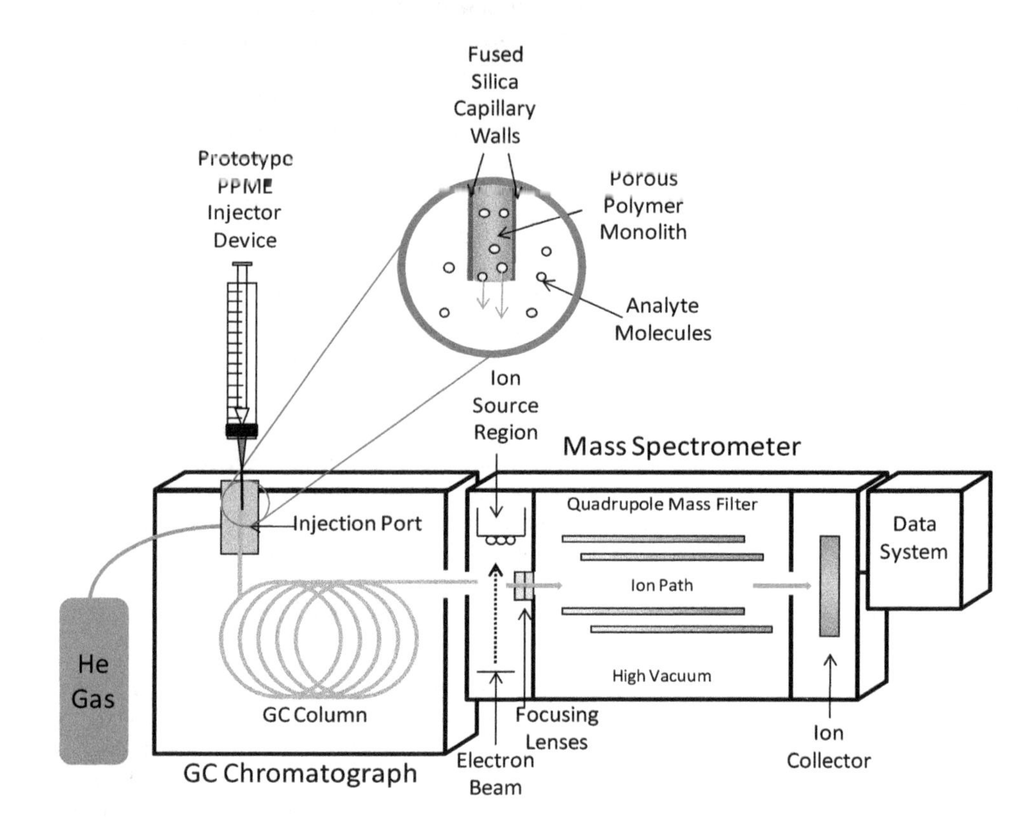

Fig. 1. Schematic of a PPME at the Injection Port of a GC-MS system.

2. Chemicals and materials

Acrylamide (99%, AA), styrene (>99%), 4-vinylpyridine (99%, 4-VP), glycidyl methacrylate (97%, GMA), ethylene glycol methacrylate (98%, EDMA), divinylbenzene (80%, DVB), 4,4'-azobis(4-cyanovaleric acid) (75+%), 1,1'-carbonyldiimidazole, 3-(trimethylsilyl) propyl methacrylate (98%), cyclohexanol (97%), 1-dodecanol (98+%), 3-aminopropyltriethoxylsilane (99%, APTES), ethanol (85%), naphthalene (≥99.7%), 2,6-dimethylnaphthalene (99%), phenanthrene (≥99.5%), HPLC grade acetonitrile, fluorene (≥99.0%), caffeine, isotopically-labelled (trimethyl ^{18}C) caffeine, megestrol acetate, 17β-estradiol, acetic acid (≥99%) and methanol were purchased from Sigma-Aldrich (St. Louis, MO, USA). Sodium acetate (99+%) was purchased from Alfa Aesar (Ward Hill, MA, USA). 99.7% NaOH and citric acid monohydrate were purchased from Fisher Scientific (Fair Lawn, NJ, USA).

The structures of analytes selected for extraction and selected reagents used for making the polymer monoliths are shown in Figure 2. All through the experiment deionized (D.I.) water used was prepared using a Millipore Elix 10 water purification system.

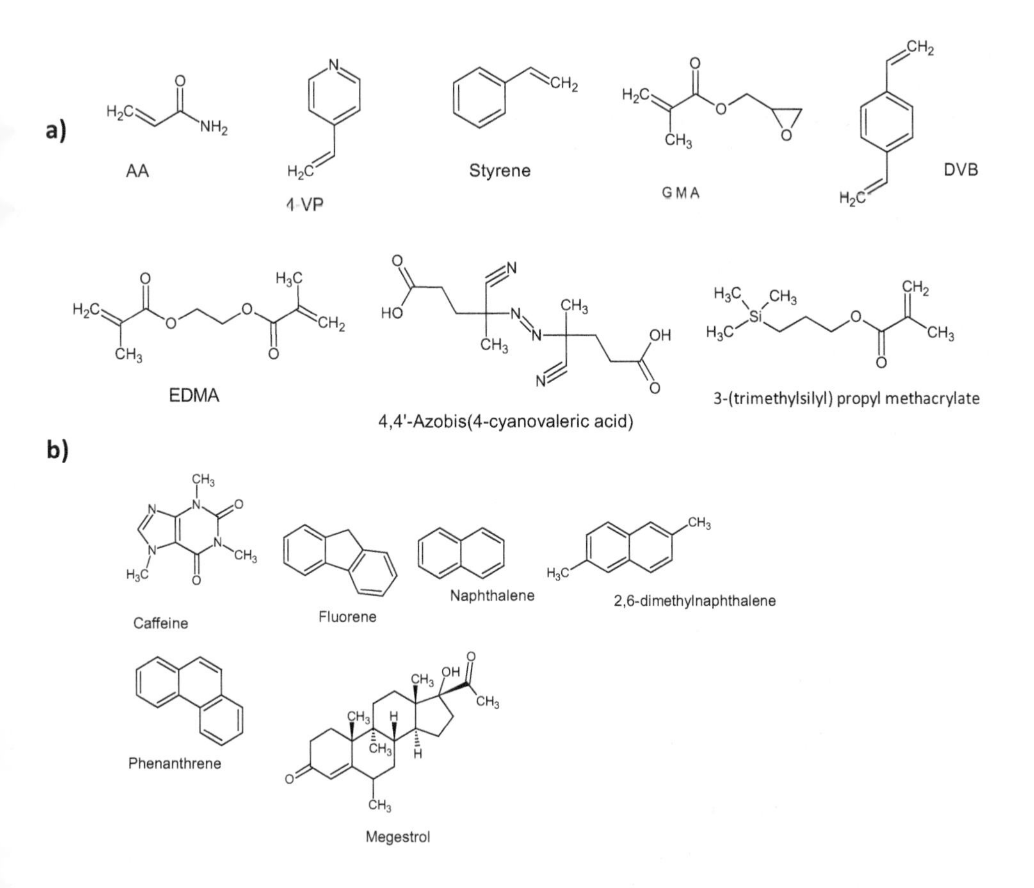

Fig. 2. a) Structures of some reagents used to make the polymer monoliths and, b) structures of some of the analytes that were tested.

3. Methods

3.1 Formation of porous polymer monoliths on fused silica capillary

3.1.1 Poly(Acrylamide-co-EDMA)

The poly(acrylamide-co-EDMA) monolithic capillary was formed inside a polyimide-coated fused silica capillary (10 cm x 250 µm, i.d., Polymicro Technologies, USA) by heat-initiated polymerization. To begin, the capillary was conditioned by flushing (using a Harvard Apparatus PHD Ultra syringe pump) it with 1.0 M NaOH at a flow rate of 10 µL/min for 3 h, 0.1 M HCl for 1 h, washed with deinionized (D.I.) water for 1 h, then dried with a flow of nitrogen. Next, the capillary was flushed at 10 μL/min with 20% (v/v) 3-(trimethoxysilyl)propyl methacrylate in toluene overnight (17 h), then infused (for 5 minutes at 5 μL/min) with the pre-polymer mixture, consisting of 13.0 mg acrylamide, 179 μL EDMA, 7 mg 4,4'-azobis(4-cyanovaleric acid) initiator, and 500 μL 1/1 (v/v) ethanol/D.I. water. Both ends of the capillary were capped and the capillary was placed in a 70°C oven overnight (16 h). Once monolith formation was complete, the capillary was flushed with 1/1 (v/v) methanol/D.I. water to remove any unreacted reagents.

3.1.2 Poly (Styrene-co-Divinylbenzene)

The poly(styrene-co-divinylbenzene) monolithic capillary was formed the same way as the preparation of poly(acrylamide-co-EDMA) monolith except the composition of prepolymer mixture, which consists of 200 μL styrene, 80 μL divinyl benzene, 600 μL ethanol, and 5 mg 4,4'-Azobis(4-cyanovaleric acid) initiator.

3.1.3 Poly (4Vinylpyridine-co-EDMA)

The poly (4Vinylpyridine-co-EDMA) monolith was formed the same as documented above for the others. The pre-polymer mixture, consisted of 179 μL of EDMA, 13 μL of 4-vinylpyridine, 7 mg of 4,4'-Azobis(4-cyanovaleric acid), 250 μL of ethanol, and 250 μL of D.I. water.

3.1.4 Poly (Glycidylmethacrylate-co-EDMA)

The poly (glycidylmethacrylate-co-EDMA) monolith was formed by polymerization of a mixture of 80 μL of EDMA, 240 μL of glycidyl methacrylate, 7 mg of 4,4'-azobis(4-cyanovaleric acid), 400 μL of cyclohexanol, and 180 μL of dodecanol.

3.1.5 Poly (styrene-co-divinylbenzene) PLOT

The poly(styrene-divinylbenzene) PLOT capillary was formed inside a 250 μm i.d., 10 cm long fused silica capillary as outlined schematically in Figure 3. This method was adapted from the literature (Schweitz, L. 2002). Briefly, the capillary was conditioned by flushing with 1.0 M NaOH for 3 h at a flow rate of 10 μL/min. This was followed by a D.I. water wash for 1 h, then dried with a flow of nitrogen. Next, the capillary was filled with 20% (v/v) APTES in toluene overnight (16 h), then rinsed with acetone for 1 h and then dried. The initiator solution, consisting of 8.4 mg of 4,4'-azobis(4-cyanovaleric acid) initiator, 9.7 mg of 1,1'-carbonyldiimidazole, 1.5 mL of methanol, and 1.5 mL of 50 mM sodium acetate-acetic acid buffer (pH ~5.5) was prepared and the pH of the final solution was adjusted to 5.5 using acetic acid. The capillary was infused at 10 μL/min with the initiator solution overnight (16 h), then washed with 1/1 (v/) methanol/D.I. water for 30 minutes and dried with a flow of nitrogen. Next, the prepolymer mixture, consisting of 200 μL styrene, 200 μL DVB, and 600 μL ethanol, was prepared and infused into the capillary for 5 min at 5 μL/min. Both ends of the capillary were capped and placed in a 70°C oven overnight (16 h) for polymerization to ensue. The PLOT column was washed with 1/1 (v/v) methanol/water.

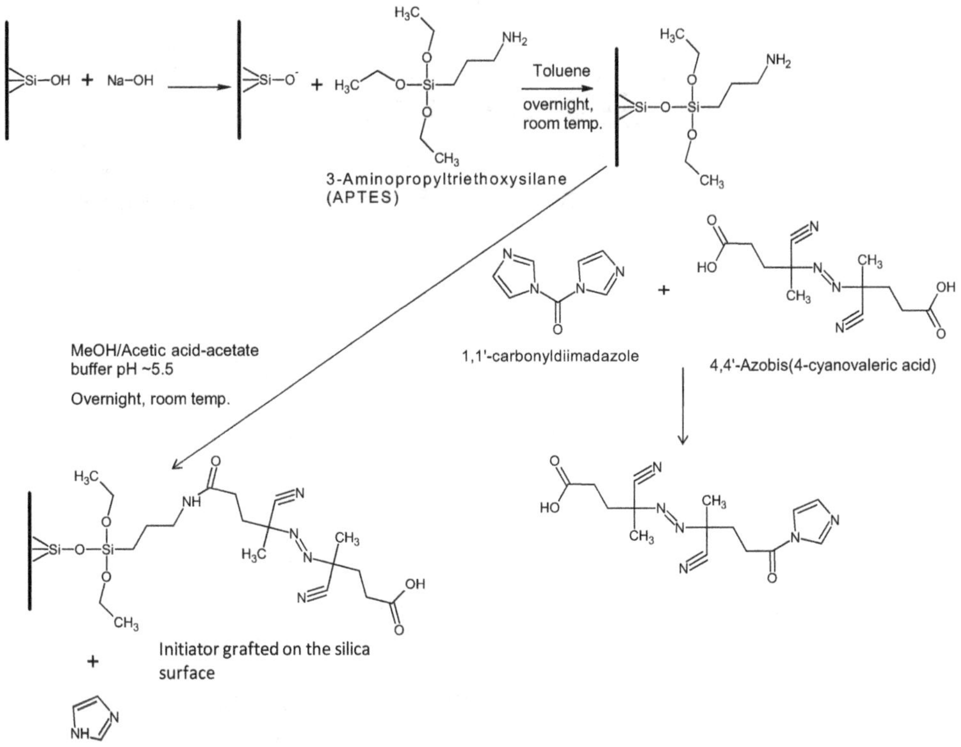

Fig. 3. Reaction schematic for grafting initiator onto the capillary wall for open tubular monolith formation

3.2 PPME device fabrication

A prototype PPME device was prepared using the porous polymer monolith described above, disposable plastic syringes with removable needles, micropipette tips, and fast-drying epoxy. A disposable plastic syringe was dismantled by removing the inner plunger. The rubber stopper was removed from the end of the plunger and the tip of the stopper was trimmed off, creating a sort of rubber washer. A 10-100 µL yellow micropipette tip was trimmed so that its wide end fits snugly over the end of the hard plastic plunger, the stem of which had previously been trimmed to fit inside the pipette tip. The pipette tip was filled with thoroughly-mixed epoxy glue then secured onto the trimmed plunger stem. A capillary containing the desired porous polymer monolith was then inserted into the narrow end of the pipette tip attached to the plunger and the rubber washer was slipped over the capillary and secured to the base of the pipette tip with epoxy. The prototype was left to dry overnight then reinserted into the syringe outer barrel. With the PPME capillary withdrawn into the syringe barrel, a needle was attached to the syringe. This enables the PPME device to puncture a GC septum effectively before the PPME capillary is injected for desorption. A schematic of the in-house fabricated PPME device is shown in Figure 4.

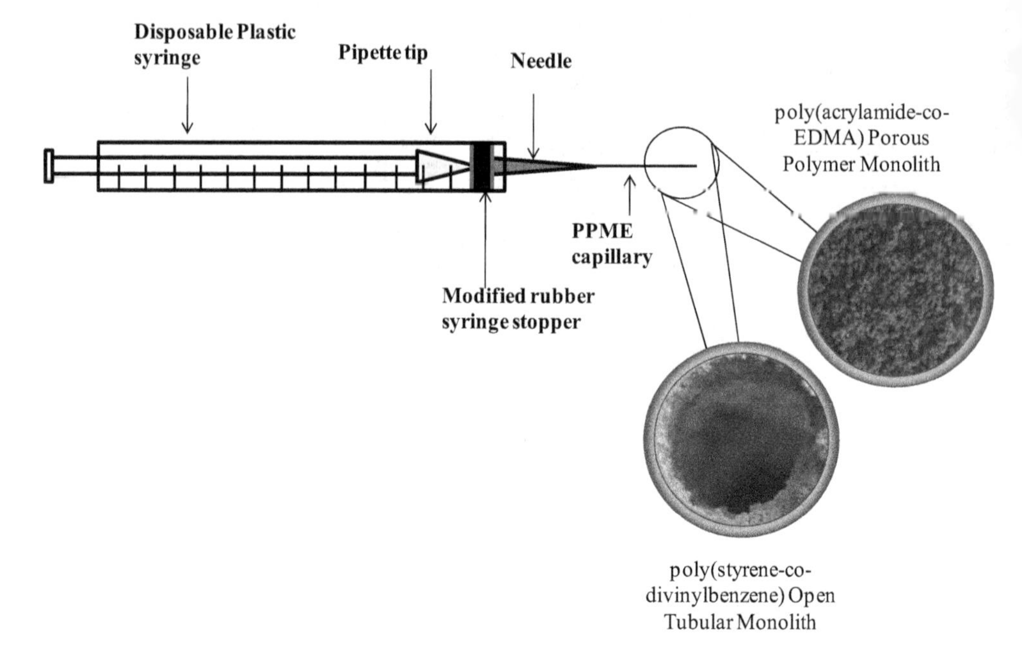

Fig. 4. A schematic of the in-house fabricated PPME device with an inset of SEM images of poly(acrylamide-co-EDMA) and poly (styrene-co-divinylbenzene) PLOT platforms.

3.3 Extraction of caffeine and PAHs using PPME and analysis by GC-MS

3.3.1 Caffeine extraction

3 mL of caffeine standards with the concentration of 100, 200, 300, and 400 µg/mL were prepared in 4 mL vials and 75 µL of the internal standard, isotopically-labelled (trimethyl ^{13}C) caffeine, was added. The poly(acrylamide-co-EDMA) PPME capillary was chosen for this extraction as it posses fairly high polarity stationary material that could preferentially adsorb caffeine. The PPME was conditioned by inserting it in the GC injection port for 30 minutes at 270°C, then cooled at room temperature. Starting with the 100 µg/mL caffeine standard, the vial was placed on a stir plate and stirred at a controlled rate. The PPME capillary was then immersed in the standard solution for 7 minutes to allow caffeine to adsorb onto the polymer monolith, then the PPME was injected into the GC port and desorbed at the GC injector for 3 minutes at 200°C. This procedure was repeated for the 200, 300, and 400 µg/mL standards with the PPME capillary being cooled at room temperature between runs. For comparison, the above procedure was also repeated using a commercially available DVB/Carboxen/PDMS StableFlex™ SPME fibre (Supelco).

3.3.2 Polyaromatic hydrocarbon (PAH) extraction

1.5 mL of PAH standards containing naphthalene, 2,6-dimethylnaphthalene, and phenanthrene with the concentration of 20, 50, 100, and 200 µg/mL were prepared in 2 mL vials with 75 µL of the internal standard fluorene (480 µg/mL) added. Based on polarity matching considerations, the poly (styrene-co-divinylbenzene) PPME device was conditioned in the GC injection port for 8 minutes at 280°C then cooled at room temperature. Starting with the 20 μg/mL standard, the vial was placed on a stir plate and stirred at a controlled rate. The PPME capillary was immersed in the standard solution for to allow the PAHs to adsorb onto the polymer monolith. The adsorption times that were evaluated included 3 minutes, 1 minute, and 30 seconds. After extraction the PPME was injected into the GC injection port (200°C) for 3 minutes' desorption. This procedure was repeated for the 50, 100, and 200 μg/mL standards with the PPME capillary being cooled at room temperature for 8 minutes between runs.

3.3.3 Evaluation of capability of the porous monoliths for hormone extraction

Hormones in the environment emerging contaminants of scientific interest. Conventional analysis approach is to extract the hormones by a C-18 SPE cartridge and elute with an appropriate solvent followed by analysis by HPLC-DAD-MS. Analysis of hormones by GC-MS often demands a derivatization process. Different derivatization reagents have been used including silylating reagent usually, N,O-bis(trimethylsilyl)trifluoroacetamide and Heptafluorobutyric acid. Our goal was to explore the use of PPME to extract the hormones and carry out on fiber derivation followed by injection of the PPME into the GC. This has been demonstrated largely on acrylate SPME fibers by several authors (Yang et al. 2006, Pan et al. 2008).

Before the attempt to carry out on-porous polymer derivatization, it was crucial first to determine which porous polymer monolith was best suited for the extraction of hormones. Megestrol standards with the concentration of 0.1, 1, 10, and 20 ppm in D.I. water were prepared. Poly(GMA-co-EDMA) and poly(4VP-co-EDMA) porous polymer monoliths were chosen to evaluate their ability to extract the hormone. Both monoliths were preconditioned by washing with 1:1 (v/v) methanol/D.I. water at 25 μL/min for 40 minutes then drying with nitrogen. Starting with the 0.1 ppm megestrol standard, each porous polymer monolith in a capillary (10 cm long) was infused with the hormone solution at 25 μL/min for 40 minutes and the eluate solution was collected in a 2 mL vial. The monoliths were then dried with nitrogen and eluted with 9:1 (v/v) methanol/D.I. water at a flow rate of 25 μL/min. The eluate was collected in a 2 mL vial. This procedure was repeated for the 1, 10, and 20 ppm hormone standards and the collected solutions were analyzed by HPLC-DAD for both the monoliths.

4. Instrumentation

4.1 GC-MS instrumentation conditions

GC analysis was carried out on an Agilent 6890N gas chromatography system with an Agilent 5975C VL MSD (Agilent Technologies, USA). Analytes were separated on a 5% phenyl methyl siloxane capillary column (30 m x 250 μm I.D., 0.25 μm film thickness, Agilent Technologies, USA). The column oven temperature was initially set at 180°C for 3 minutes and then was increased to 205°C at 5°C/min. This temperature was held till the end of the analysis. The MS detector temperature was set at 200°C. The injection port temperature was set at 280°C with a split-less injection mode. Helium was used as carrier gas set at the flow rate of 24.1 mL/min.

4.2 HPLC -DAD Instrumentation and conditions

Analysis of megestrol eluted from the porous polymer monolith were performed on a Thermo LCQ fleet HPLC-DAD system consisting of a LC pump, an autosampler, a 6-port injection valve with a 20-µL injection loop, and a diode array detector (DAD) with the scan range from 250 to 400 nm. The temperature of column oven was set at 30 °C. The temperature of autosampler tray was set at 6 °C to prevent degradation of the hormones. An Agilent Zorbox Eclipse PAH RPLC column (250 x 3.0 mm I.D., 5 µm) (Agilent, Santa Clara, CA, USA) was used for the separation of hormone. The mobile phase used was acetonitrile (C) and 0.1% formic acid in H_2O (D) at a flow rate of 500.0 µL/min. The eluent gradient setting is shown in Table 1.

Time(min) \ %M.P.	Mobile phase C %	Mobile phase D %
0	20	80
5	55	45
6	95	5
10	95	5
12	20	80
15	20	80

Table 1. HPLC gradient setting for the separation of antibiotics/hormones.

5. Results and discussion

5.1 Monolithic structure of PPME and PLOT

The structure of different PPME and PLOT was investigated by scanning electron microscope (SEM), shown in Figure 3. As can be seen in Figure 3 (inset picture consist of SEM image of poly(acrylamide-co-EDMA) and poly (styrene-co-DVB) PLOT), there were many macropores and flow-through channels in the network skeleton , which can provide the high porosity and good mass transfer desired for a microextraction platform. The SEM image of poly (styrene-DVB) PLOT confirms that using the initiator capillary wall grafting procedure outlined in Figure 3, the polymer monolith only formed around the walls of the capillary. Using this procedure many other different types of open tubular polymer monoliths can be fabricated. It is proposed that the PLOT structure could potentially result in a faster mass transfer and efficient analyte extraction compared to the conventional polymer monolith. At present, research is still being conducted to confirm and test this hypothesis.

5.2 Applications of PPME

5.2.1 Caffeine extraction

Different concentrations of caffeine standards from 100 to 400 µg/mL, with isotopically-labelled (trimethyl ^{13}C) caffeine was used as the internal standard, were extracted using a poly (acrylamide-co-EDMA) PPME. The calibration curve obtained is shown in Figure 5 a). The y-axis on the calibration curve is the peak area ratio of caffeine (m/z 194) to the internal standard (m/z 197), obtained from extracted ions chromatogram. The linear regression coefficient (R^2) achieved with the PPME was 0.9771. This satisfactory R^2 indicates that this in-house built poly (acrylamide-co-EDMA) PPME is a suitable analytical tool for extracting caffeine in a wide range of concentration. We compared the extraction efficiency of the in-house fabricated PPME with the commercial SPME (2 cm-50/30μm DVB/Carboxen™/PDMS StableFlex ™ purchased from Supelco (Bellefonte, PA, USA)). The linear regression coefficient for the commercial SPME was 0.999 (as shown in Figure 5 b), which certainly rivals the in-house made PPME. This could partly be due to potential better mass transfer for the commercial SPME due to the thin film thickness (30 μm) of the stationary phase. However, it is expected by fabricating open tubular polymer monolith based PPME the mass transfer effect could improve and the results could compare with the commercial one. However, the results obtained with the PPME herein shows good usability especially with possible benefits over the commercial SPME, such as better sample capacity and less prone to breakage and thus can be reused for a much longer time.

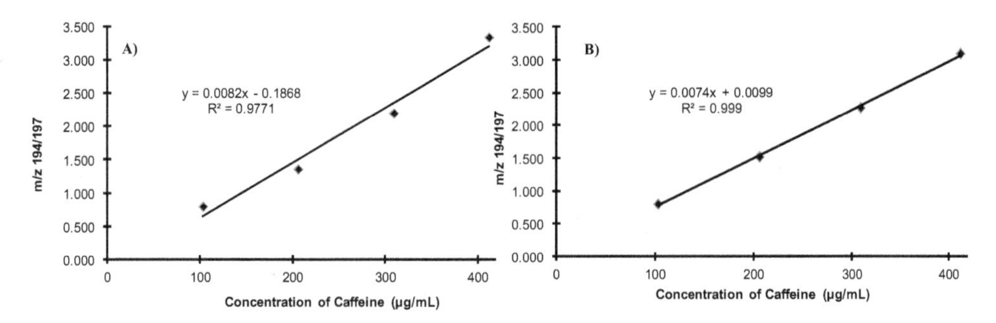

Fig. 5. Calibration curve of caffeine extracted using a) Poly (acrylamide-co-EDMA) PPME and b) commercial SPME.

The successful extraction of caffeine by PPME suggests an obvious extension of this extraction method to analysis of other compounds such as, theobromine, and theophylline and other compounds in the methylxanthine class. Theobromine is found in chocolate, tea leaves, and the cola nut, while theophylline is found in cocoa beans as well as in therapeutic drugs for respiratory disorders. Because of their presence in food and drugs it is clear that an effective and efficient extraction method for methylxanthines would be a relevant application. Using a multi-step extraction, which included the use a ^{18}C extraction tube SPE followed by filtration, Aresta's group reported the determination of caffeine, theobromine, and theophylline, among other compounds, in human milk by HPLC-UV (Aresta & Zambonin, 2005). Since it has been shown that PPME is a viable method for caffeine extraction, it is reasonable to suggest that PPME may also be used to analyze other methylxanthines, and other polar like compounds of interest in pharmaceutical, nutraceutical and environmental fields. Moreover, PPME may be more efficient, reusable, could reduce analysis cost by reducing the steps needed in analysis process and in reducing the solvents required.

5.2.2 Extraction of polyaromatic hydrocarbons

With the successful extraction of caffeine using PPME, the extraction of another class of analytes, polyaromatic hydrocarbons (PAHs) using poly(styrene-co-DVB) PPME (a fairly non polar material whose polarity could potentially march the PAHs) were also attempted. A mixture of four different PAHs at different concentrations were tested. The PAHs tested included naphthalene, 2,6-dimethylnaphthalene, phenanthrene and florene, where florene was also used as an internal standard. While good results were obtained at 3 min, 1 min adsorption time, it was found that even with a 30-second adsorption, satisfactory extraction of the PAHs was still achieved. Figure 6 showed the GC-MS chromatogram of four PAHs extracted by poly(styrene-co-DVB) PPME with 30- second adsorption time. Good baseline separation was achieved as attested in the chromatogram, with enough resolution to allow separation of more PAHs. This potential is important in environmental samples which can have numerous PAHs present. The PAHs peaks on the PPME showed some tailing, possibly attributable to analytes slow mass transfer. We propose this could be alleviated by the use of a PPME fabricated poly(styrene-co-DVB) PLOT platform. This work is in progress.

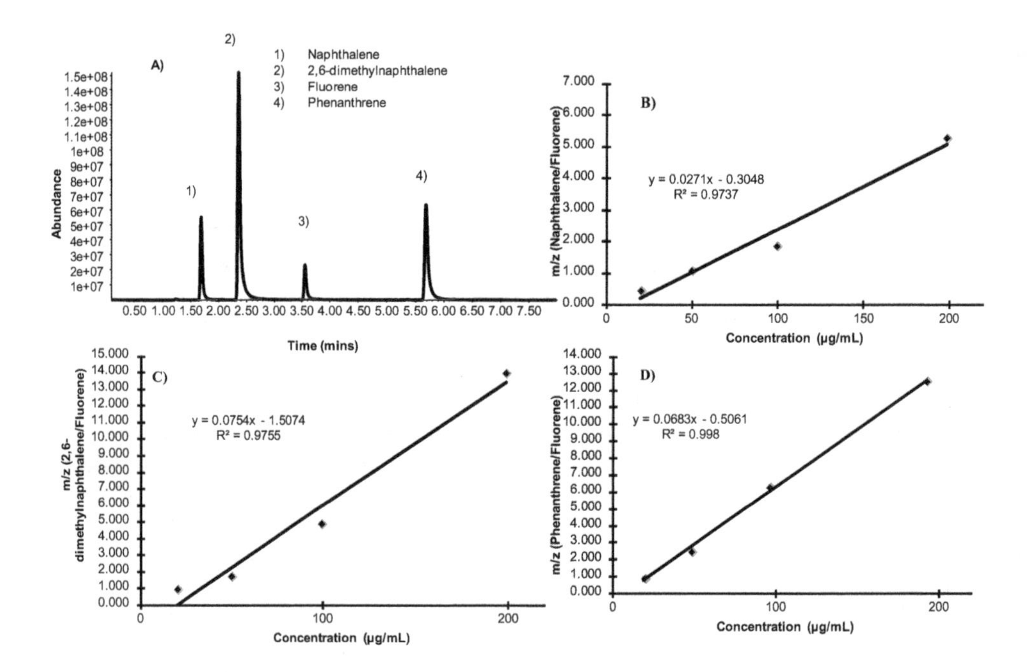

Fig. 6. A) GC-MS chromatogram showing separation of 100 µg/mL of three PAHs including; 1) naphthalene, 2) 2,6-dimethylnaphthalene, 3) fluorene, and 4) phenanthrene extracted by poly(Styrene-co-DVB) PPME; Calibration curves of B)naphthalene; C) 2,6-dimethylnaphthalene; D) phenanthrene extracted by poly(Styrene-co-DVB) PPME with 30 second adsorption and fluorene as an internal standard.

Calibration curves of the three PAHs extracted by poly(Styrene-co-DVB) PPME are shown in Figure 7 b,c,d, with analytes concentration range from 0 to 200 µg/mL. All the three PAHs showed satisfactory linear calibration curves with R^2 higher than 0.997 (R^2 = 0.9737 for naphthanlene, 0.9755 for 2,6-dimethylnaphthalene and 0.998 for phenanthrene). Clearly, the results obtained using the in-house fabricated poly (styrene-co-DVB) PPME revealed several things. First, the well-defined peaks in chromatograms obtained from GC-MS analysis showed that PAHs, even in low concentrations, are readily extracted and desorbed from the PPME capillary, making PPME a suitable method for PAH extraction. Second, this method showed feasibility for environmental monitoring of PAHs in bodies of water such as lakes, rivers, and ponds. The PPME, being more robust and rugged than conventional SPME fibers, can be taken to monitoring sites, which could make sampling process much easier. Third, this method can be extended to the analysis of any hydrophobic anyalytes immersed in an aqueous solution, since the analytes would very readily adsorb onto the hydrophobic polymer . With this knowledge, a plethora of polymers can be fabricated, each with a specific targeted analytes class to be extracted from a given sample matrix.

5.2.3 Extraction efficiency of megestrol by the polymer monoliths

There is growing public concern for the presence of hormones in the ecosystem, and the possible pathways by which these substances can enter into human and animal food chains. The concern focuses particularly on the common use of these substances in animal husbandry especially as growth promoting agents. Due to their stability most of the hormones are excreted unchanged or as active metabolites. Most commonly used processes are based on the extraction of hormones from a sample matrix and analysis by liquid chromatography (Mitani et al., 2005; Yi et al., 2007). Analysis by GC-MS could however be faster, but the hormones require derivatization to make them more volatile. This has been actualized typically by extraction of the hormones from sample matrix by C-18 cartridges, followed by pre-concentration, derivatization and then GC-MS analysis. A new method called on-fiber derivatization allows the hormones to be extracted, derivatized on the solid-phase and subsequently analyzed directly by GC-MS. Carpinteiro and coworkers combined SPME and on-fiber derivatization to detect and quantify estrogens in water by GC-MS. Samples were first extracted by SPME, and then were silylated on the headspace of a vial containing the derivatization agent. The MS signal intensity were dramatically increased by the derivatization step. The study of matrix effects demonstrated that the method is applicable for the determination of estrogens in both surface and sewage water (Carpinteiro et al., 2004). This on-fiber derivatization method eliminates the further sample preparation step before injecting into GC-MS. We propose PPME could have even better versatility in carrying out extraction and on-porous polymer derivatization of the hormones.

To test the hypothesis, two types of monoliths, poly(4VP-co-EDMA) and poly (GMA-co-EDMA) were first tested for their extraction efficiency of megestrol. We evaluated the % recovery of megestrol using HPLC-DAD analysis. Both poly(4VP-co-EDMA) and poly (GMA-co-EDMA) were found to have very good adsorptive capacity for megestrol with calculated % recovery of 98% and 107% respectively. An overlapped chromatogram indicating megestrol peak for poly(4VP-co-EDMA) , poly (GMA-co-EDMA) and directly injected 10 ppm megestrol standard is shown in Figure 8. These two polymers were found to be good stationary phases for analysis of hormones.

At the point of publication of this chapter, the demonstration of the on-fiber derivatization of the hormones using PPME fabricated from poly(4VP-co-EDMA) and poly (GMA-co-EDMA) was ongoing.

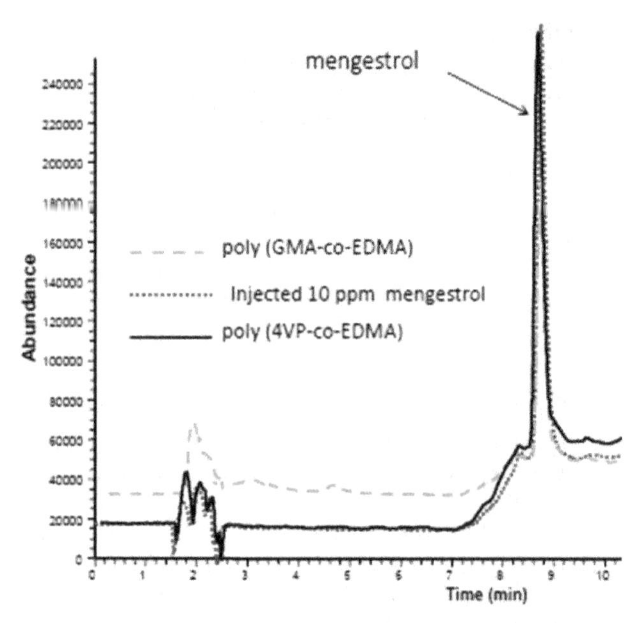

Fig. 8. Overlapped chromatogram indicating megestrol peak for poly(4VP-co-EDMA) and poly (GMA-co-EDMA) and directly injected 10 ppm megestrol standard.

6. Conclusions

Polymer monolith microextraction (PPME) has been demonstrated to be feasible, facile and versatile platform for analysis, preconcentration and extraction of analytes and for online coupling to GC-MS. The monoliths that have been tested including poly (styrene-co-divinylbenzene), poly(acrylamide-co-EDMA), poly (4-vinylpyridine-co-EDMA) and poly (glycidyl methacrylate-co-EDMA, have been found to work well when matched with analytes of corresponding polarity. The polymer monoliths based materials provided for satisfactorily fast mass transfer and thus good extraction efficiency. Good linear calibration curves were achieved in a wide concentration range when using poly(acrylamide-co-EDMA) for analysis of caffeine, poly (styrene-co-divinylbenzene) for analysis of PAHs. Both poly (4-vinylpyridine-co-EDMA) and poly (glycidyl methacrylate-co-EDMA were tested and found to be good sorbents for extraction of hormones such as megestrol with good % recoveries of 98% and 107% respectively, which suggests the potential of applying this PPME method to a wider selection of hormones and to a more complicated sample matrix.

An analogous PLOT polymer monolith was successively prepared and is being tested as a microextraction platform.

7. References

Aresta, A.; Palmisano, F. & Zambonin, C. G. (2005). Simultaneous determination of caffeine, theobromine, *Food Chemistry*, Vol. 93, Issue 1, (November 2005), pp. 177-181, ISSN 0308-8146.

Arthur, C.L. & Pawliszyn, J. (1990). Solid phase microextraction with thermal desorption using fused silica optical fibers. *Analytical Chemistry*, Vol. 62, No. 19, (October 1990), pp. 2145–2148, ISSN 0003-2700.

Bravo, J. C.; Fernández, P. & Dúrand, J. S. (2005). Flow injection fluorimetric determination of β-estradiol using a molecularly imprinted polymer. *Analyst*, Vol. 130, No.10, (August 2005), pp. 1404-1409, ISSN 0003-2654.

Carpinteiro, J.; Quintana, J.B.; Rodriguez, I.; Carro, A.M.; Lorenzo, R.A. & Cela, R. (2004). Applicability of solid-phase microextraction followed by on-fiber silylation for the determination of estrogens in water samples by gas chromatography-tandem mass spectrometry. *Journal of Chromatography A*, Vol. 1056, No.1-2, (November 2004), pp. 179-185, ISSN 0021-9673.

Fan, Y.; Zhang, M. & Feng, Y. Q. (2005). Poly (acrylamide-vinylpyridine-N, N'-methylene bisacrylamide) monolithic capillary for in-tube solid-phase microextraction coupled to high performance liquid chromatography. *Journal of Chromatography A*, Vol. 1099, No. 1-2, (September 2005), pp. 84-91, ISSN 0021-9673.

Gibson, G.T.; Mugo, S.M.& Oleschuk, R.D. (2008). Surface-mediated effects on porous polymer monolith formation within capillaries. *Polymer*, vol. 49, No. 13-14, (June 2008), pp.3084–3090, ISSN 0032-3861

Görög, S.(2011). Advances in the analysis of steroid hormone drugs in pharmaceuticals and environmental samples. *Journal of Pharmaceutical and Biomedical Analysis*, Vol. 55, No. 4, (November 2010), pp.728–743, ISSN 0731-7085.

Jeannot, M. A. & Cantwell, F. F. (1996). Solvent Microextraction into a Single Drop. *Analytical Chemistry*, Vol. 68, No. 13, (July 1996), pp. 2236-2240, ISSN 0003-2700.

Körner, W.; Bolz, U.; Süßmuth, W.; Hiller, G.; Schuller, W.; Hanf, V. & Hagenmaier, H. (2000). Input/output balance of estrogenic active compounds in a major municipal sewage plant in Germany. *Chemosphere*, Vol. 40, No. 9-11, (May 2000), pp. 1131-1142, ISSN 0045-6535.

Liu, L.; Cheng, J.; Matsadiq, G. & Li, J. K. (2011). Novel polymer monolith microextraction using a poly-(methyl methacrylate-co-ethylene dimethacrylate) monolith and its application to the determination of polychlorinated biphenyls in water samples to the determination of polychlorinated biphenyls in water samples. *Chemosphere*, Vol. 83, No. 10, (April 2011), pp. 1307-1312, ISSN 0045-6535.

Mitani, K.; Fujioka, M. & Kataoka, H. (2005). Fully automated analysis of estrogens in environmental waters by in-tube solid-phase microextraction coupled with liquid chromatography-tandem mass spectrometry. *Journal of Chromatography A*, Vol. 1081, No. 2, (July 2005), pp. 218-224, ISSN 0021-9673.

Pan, Y.P. & Tsai, S. W. (2008). Solid phase microextraction procedure for the determination of alkylphenols in water by on-fiber derivatization with N-tert-butyl-dimethylsilyl-N-methyltrifluoroacetimide. *Analytica Chimica Acta*, Vol. 624, No. 2, (August 2008), pp. 247-252, ISSN 0003-2670.

Park, C. H.; Park, S. J.& Shin, H. S. (2001). Sensitive determination of bisphenol A in environmental water by gas chromatography with nitrogen–phosphorus detection after cyano methylation. *Journal of Chromatography A*, Vol.912, No.1, (March 2001), pp. 119-125, ISSN 0021- 9673.

Planeta, J.; Moravcová, D.; Roth, M.; Karásek, P.& Kahle, V. (2010). Silica-based monolithic capillary columns- Effect of preparation temperature on separation efficiency.

Journal of Chromatography A, Vol. 1217, No. 36, (July 2010), pp. 5737-5740, ISSN 0021-9673.

Pena-Pereira, F.; Lavilla, I.& Bendicho, C. (2009). Miniaturized preconcentration methods based on liquid-liquid extraction and their application in inorganic ultratrace analysis and speciation: A review. *Spectrochimica Acta Part B: Atomic Spectroscopy*, Vol. 64, No. 1, (January 2009), pp. 1-15, ISSN 0584-8547.

Sambe, H.& Haginaka, J. (2003) Uniformly sized molecularly imprinted polymers for bisphenol A and β-estradiol; retention and molecular recognition properties in hydro-organic mobile phases. *Journal of Pharmaceutical and Biomedical Analysis*, Vol. 30, No. 6, (2003), pp. 1835-1844, ISSN 0731-7085.

Schweitz, L. (2002). Molecularly imprinted polymer coatings for open-tubular capillary electrochromatography prepared by surface initiation. *Analytical Chemistry*, Vol. 74, No.5, (January 2002), pp. 1192-1196, ISSN 0003-2700.

Svec, F. (2010) Porous polymer monoliths: amazingly wide variety of techniques enabling their preparation. *Journal of Chromatography A*, Vol. 1217, No. 6, (February 2010), pp. 902-924, ISSN 0021-9673.

Ueta, I.; Saito, Y.; Teraoka. K.; Matsuura, H.; Fujimura, K.& Jinno, K. (2010). Novel Fire Investigation Technique Using Needle Extraction in Gas Chromatography. *Analytical Sciences*, Vol. 26, No. 11, (November 2010), pp. 1127-1132, ISSN 0910-6340.

Wen, Y.; Zhou, B. S.; Xu, Y.; Jin, S. W. & Feng, Y. Q. (2006). Analysis of estrogens in environmental waters using polymer monolith in polyether ether ketone tube solid-phase microextraction combined with high-performance liquid chromatography. *Journal of Chromatography A*, Vol. 1133, No. 1-2, (November 2006), pp. 21-28, ISSN 0021-9673.

Yang, L.; Lan, C.; Liu,H.; Dong, J. & Luan, T. (2006). Full automation of solid-phase microextraction/on-fiber derivatization for simultaneous determination of endocrine-disrupting chemicals and steroid hormones by gas chromatography–mass spectrometry. *Analytical and Bioanalytical Chemistry*, Vol. 386, No.2, (July 2006), pp. 391-397, ISSN 1618-2642.

Yi, W.; Ying, W.; Bing-Sheng, Z.; Xu, Y. & Yu-Qi, F. (2007). Determination of sexual hormones in liquid cosmetics by polymer monolith microextraction coupled with high performance liquid chromatography. *Chinese Journal of Analytical Chemistry*, Vol. 35, No. 5, (May 2007), pp. 681-684, ISSN 1872-2040.

Zhang, M.; Fang, W.; Zhang, Y.F.; Nie, J. & Feng, Y.Q. (2006). Novel polymer monolith microextraction using a poly (methacrylic acid-ethylene glycol dimethacrylate) monolith and its application to simultaneous analysis of several angiotensin II receptor antagonists in human urine by capillary zone electrophoresis. *Journal of Chromatography A*, Vol. 1102, No. 1-2, (January 2006), pp. 294–301, ISSN 0021-9673.

Zhang, Z. & Pawliszyn, J. (1995). Quantitative extraction using an internally cooled solid phase microextraction device. Analytical Chemistry, Vol. 67, No. 1, , pp. 34-43, ISSN 0003-2700.

Zhou, J.; Ma, C.; Zhou, S.; Ma, P.; Chen F.; Qi, Y. & Chen, H. (2010). Preparation, evaluation and application of molecularly imprinted solid-phase microextraction monolith for selective extraction of pirimicarb in tomato and pear. *Journal of Chromatography A*, Vol. 1217, No. 48, (November 2010), pp. 7478-7483, ISSN 0021-9673.

Parameters Influencing on Sensitivities of Polycyclic Aromatic Hydrocarbons Measured by Shimadzu GCMS-QP2010 Ultra

S. Pongpiachan[1,2,*], P. Hirunyatrakul[3], I. Kittikoon[3] and C. Khumsup[3]
[1]NIDA Center for Research and Development on Disaster Prevention and Management, School of Social and Environmental Development, National Institute of Development, Administration (NIDA), Sereethai Road, Klong-Chan, Bangkapi, Bangkok,
[2]SKLLQG, Institute of Earth Environment, Chinese Academy of Sciences (IEECAS), Xi'an,
[3]Bara Scientific Co., Ltd., Bangkok,
[1,3]Thailand
[2]China

1. Introduction

Throughout the decades of analytical instrument designs and developments there have been many incidents and scientific improvements, which have assisted to generate major design trend changes. A unique and innovative technology enabled analytical instrument designed to meet the needs of quantitative chemical analysis of polycyclic aromatic hydrocarbons (PAHs) in various environmental compartments. PAHs are a class of very stable organic molecules made up of only carbon and hydrogen and contain two to eight fused aromatic rings. PAHs are formed during incomplete combustion of organic materials such as fossil fuels, coke and wood. These molecules were oriented horizontal to the surface, with each carbon having three neighboring atoms much like graphite. The physicochemical properties and structures of a variety of representative PAHs can be seen in Table 1-2. Epidemiological evidence suggests that human exposures to PAHs, especially Benzo[a]pyrene are high risk factors for carcinogenic and mutagenic effects. There are hundreds of PAH compounds in the environment, but only 16 of them are included in the priority pollutants list of US EPA (EPA, 2003). Many PAHs have also been identified as cancer-inducing chemicals for animals and/or humans (IARC, 1983).

In 1775, the British surgeon, Percival Pott, was the first to consider PAHs as toxic chemicals with the high incidence of scrotal cancer in chimney sweep apprentices (IARC, 1985). Occupational exposure of workers by inhalation of PAHs, both volatile and bound to respirable particulate matter, and by dermal contact with PAH-containing materials, occurs at high levels during coke production, coal gasification, and iron and steel founding. Coke oven workers have a 3- to 7- fold risk increase for developing lung cancer (IARC, 1984 and IARC, 1987).

* Corresponding Author

Congeners	MW (g/mol)	MP (°C)	BP (°C)	P_S	P_L	Log K_{ow}	H	Log K_{OA}
Ace	154.2	96	277.5	0.3	1.52	3.92	12.17	6.23
Ac	150.2	92	265-275	0.9	4.14	4.00	8.40	6.47
Fl	166.2	116	295	0.09	0.72; 0.79[a]	4.18	7.87	6.68
Ph	178.2	101	339	0.02	0.11; 0.06	4.57	3.24	7.45
1-MePh	192.3	123	359			5.14		
An	178.2	216	340	1.00E-03	7.78E-02	4.54	3.96	7.34
Pyr	202.3	156	360	6.00E-04	1.19E-02; 8E-03	5.18	0.92	8.61
Flu	202.3	111	375	1.23E-03	8.72E-03	5.22	1.04	8.60
B[*a*]F	216.3	187	407			5.40		
B[*b*]F	216.3	209	402			5.75		
Chry	228.3	255	448	5.70E-07	1.07E-04	5.86	6.50E-02	10.44
Tri	228.3	199	438	2.30E-06	1.21E-04	5.49	1.20E-02	10.80
p-terp	230.1	213		4.86E-06		6.03		
B[*a*]A	228.3	160	435	2.80E-05	6.06E-04	5.91	0.58	9.54
B[*a*]P	252.3	175	495	7.00E-07	2.13E-05	6.04	4.60E-02	10.77
B[*e*]P	252.3	178		7.40E-07	2.41E-05		0.02	
Per	252.3	277	495	1.40E-08		6.25	3.00E-03	12.17
B[*b*]F	252.3	168	481			5.80		
B[*j*]F	252.3	166	480					
B[*k*]F	252.3	217	481	5.20E-08	4.12E-06	6.00	1.60E-02	11.19
B[*g,h,i*]P	268.4	277			2.25E-05	6.50	7.50E-02	11.02
D[*a,h*]A	278.4	267	524	3.70E-10	9.16E-08	6.75		
Cor	300.4	>350	525	2.00E-10		6.75		

Source: http://www.es.lancs.ac.uk/ecerg/kcjgroup/5.html
MP (°C): Melting Point, K_{ow} : Octanol-water partition coefficient, BP (°C): Boiling Point,
H: Henry 's Law Constant, P_S : Vapour pressure of solid substance
K_{oa}: Octanol-air partition coefficient, P_L: Vapour pressure of subcooled liquid

Table 1. Physiochemical properties of PAHs

For this reason, the monitoring of PAHs in environmental media is a reasonable approach to assess the risk for adverse health effects. Since the fate of PAHs in the natural environment is mainly governed by its physiochemical properties, the study of general properties of the compounds is of great concern. It is well known that aqueous solubility, volatility (e.g. Henry's law constant of air/water partition coefficient, octanol/air partition coefficient), hydrophobicity or lipophilicity (e.g., n-octanol/water partition coefficient) of PAHs vary widely (Mackay and Callcot, 1998), the differences among their distribution in aquatic systems, the atmosphere, and the soil are significant. Molecular weight and chemical structure influence physical and chemical properties between individual PAHs. The vapor pressure and water solubility basically decrease with the increasing molecular weight. The fate of PAHs in the environment is largely determined by its physiochemical properties; as a result, high mobility of low molecular weight (LMW) species can be expected (Wild and Jones, 1995).

Congener	Abbreviation	M.W. [g]	Chemical Structure
Acenaphthylene	Ac	152	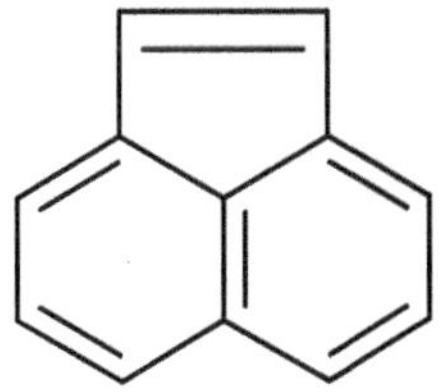
Acenaphthene	Ace	154	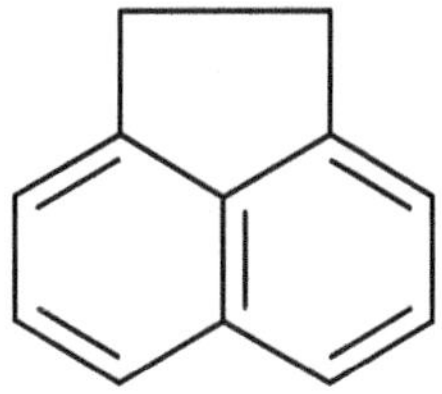
Fluorene	Fl	166	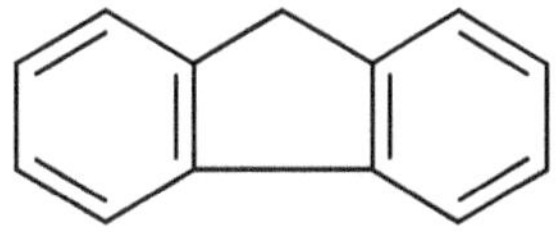
Phenanthrene	Ph	178	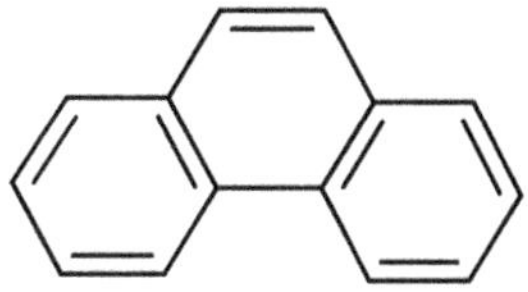
Anthracene	An	178	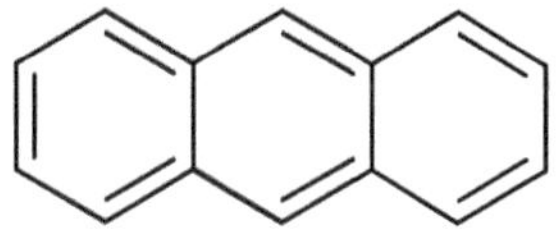
3-Methyl Phenanthrene	3-MePh	192	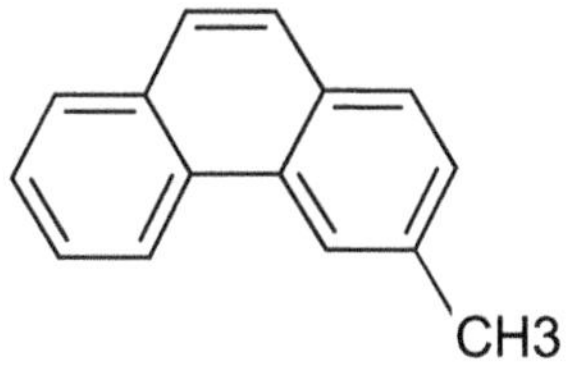 CH3

Congener	Abbreviation	M.W. [g]	Chemical Structure
9-Methyl Phenanthrene	9-MePh	192	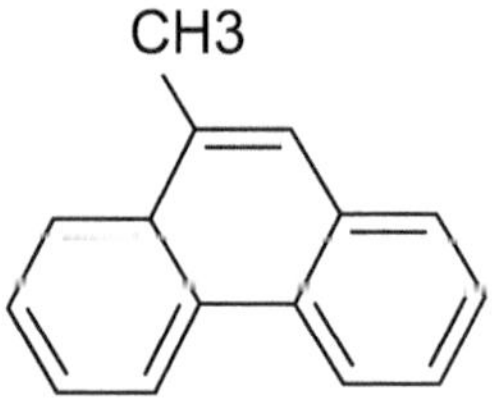 CH3
1-Methyl Phenanthrene	1-MePh	192	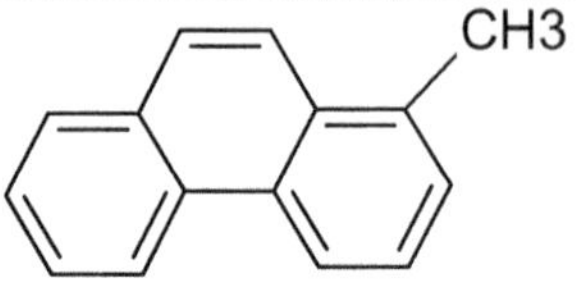 CH3
2-Methyl Phenanthrene	2-MePh	192	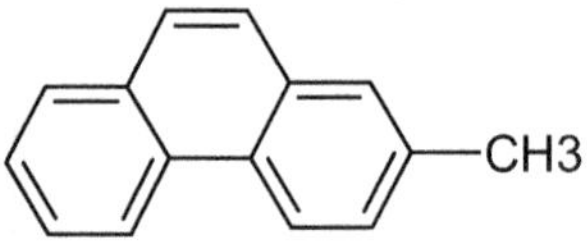 CH3
1-methyl-7-isopropyl phenanthrene (Retene)	Ret	234	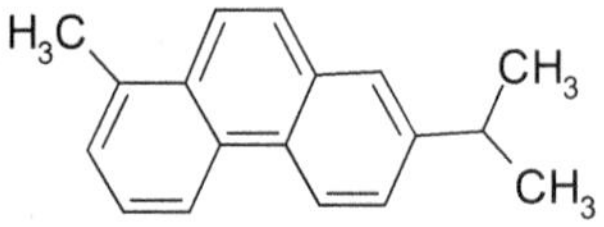 H_3C CH_3 CH_3
Fluoranthene	Fluo	202	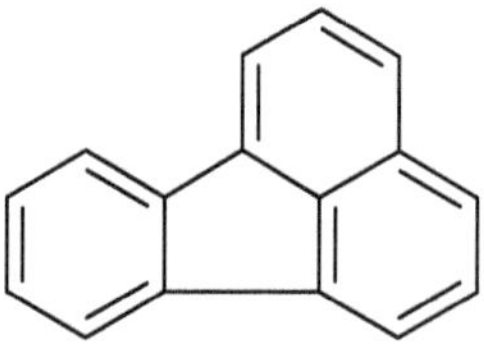
Pyrene	Py	202	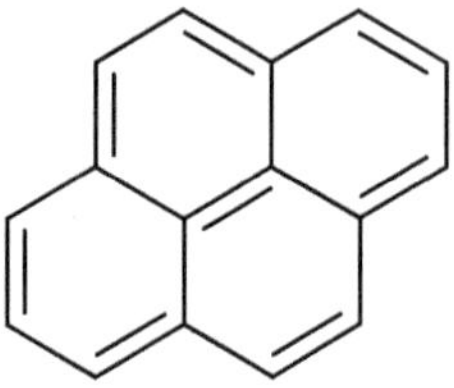
Benz[*a*]anthracene	B[*a*]A	228	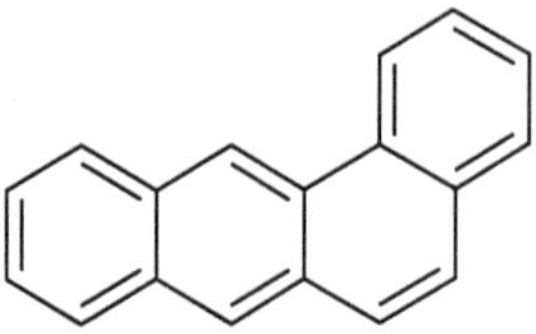

Congener	Abbreviation	M.W. [g]	Chemical Structure
Chrysene	Chry	228	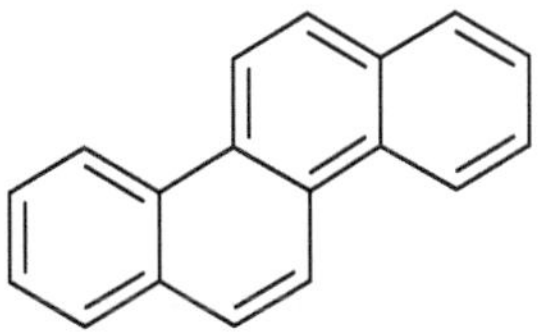
Triphenylene	Tri	228	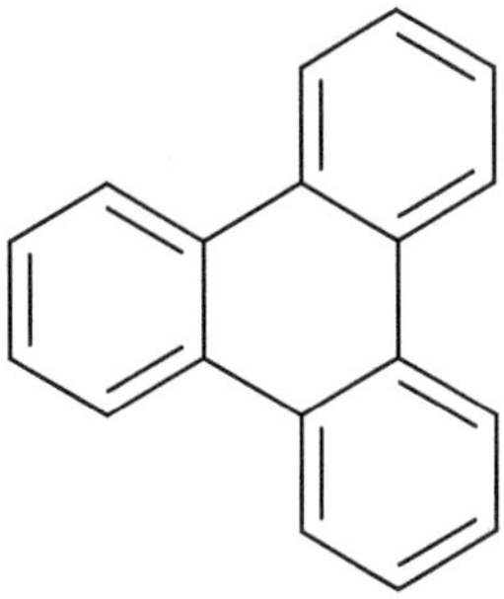
Benzo[*b*]fluoranthene	B[*b*]F	252	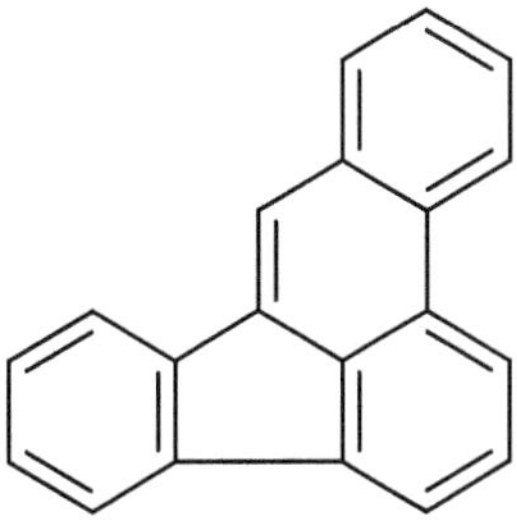
Benzo[*j*]fluoranthene	B[*j*]F	252	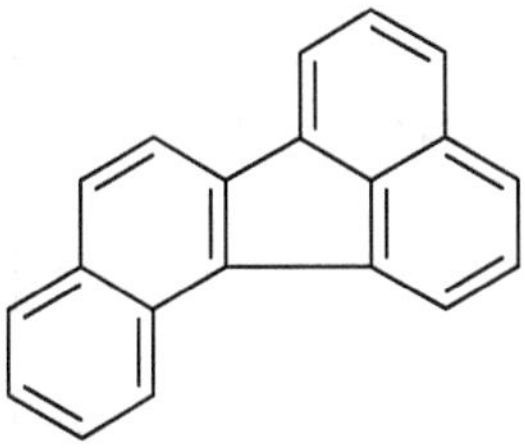
Benzo[*k*]fluoranthene	B[*k*]F	252	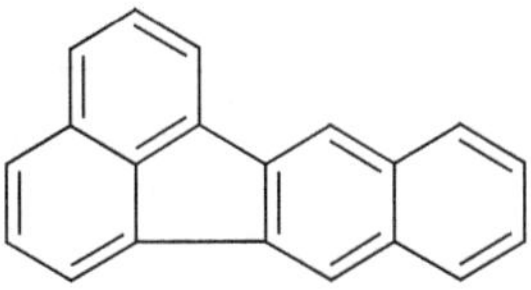

Congener	Abbreviation	M.W. [g]	Chemical Structure
Benzo[*e*]pyrene	B[*e*]P	252	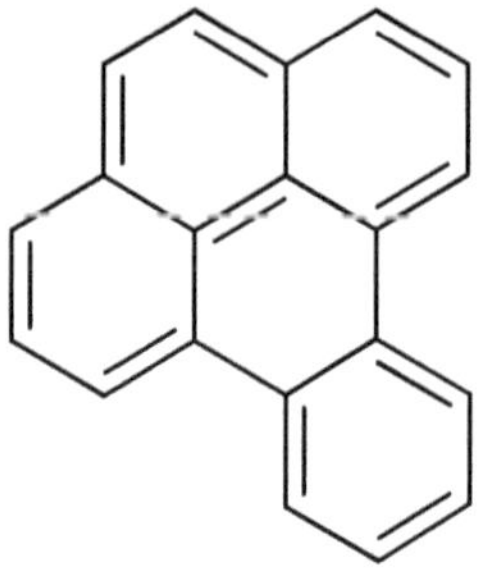
Benzo[*a*]pyrene	B[*a*]P	252	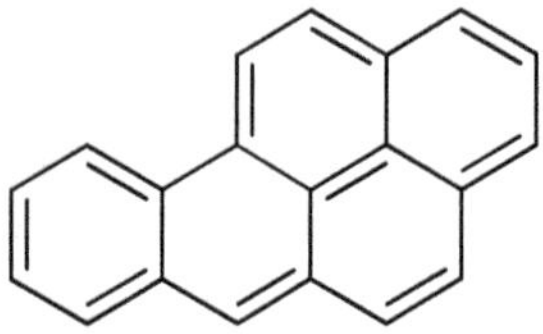
Perylene	Per	252	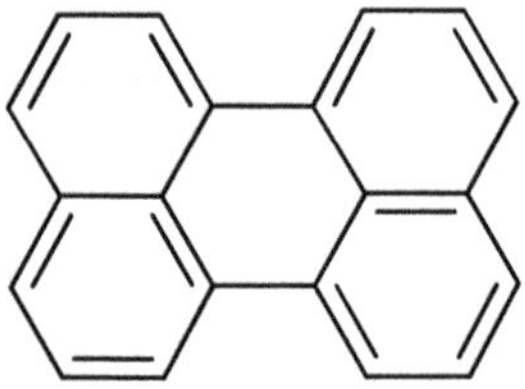
Indeno[*1,2,3-cd*]pyrene	Ind	276	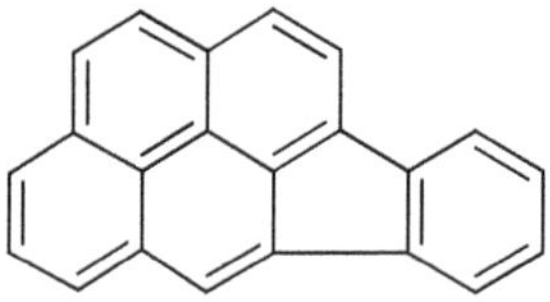
Benzo[*g,h,i*]perylene	B[*g,h,i*]P	276	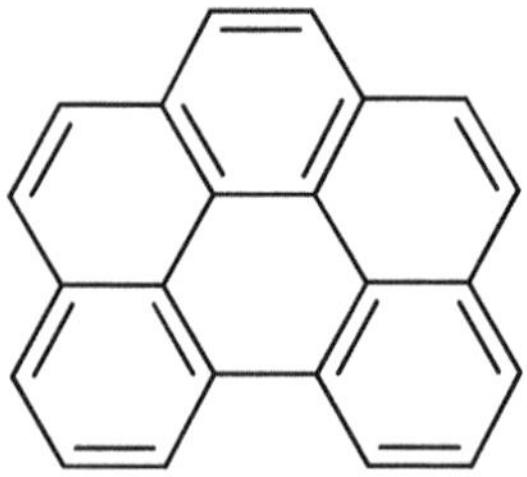
Anthanthrene	Ant	276	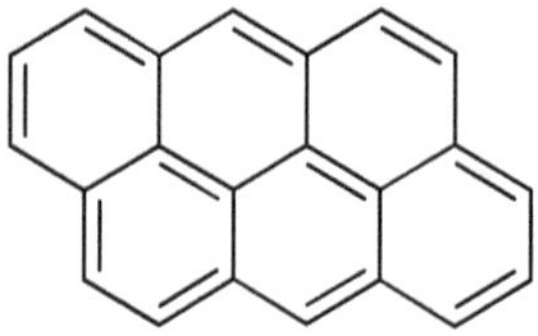

Congener	Abbreviation	M.W. [g]	Chemical Structure
Dibenzo[*a,h*]anthracene	D[*a,h*]A	278	
Coronene	Cor	300	

Table 2. Chemical structures of PAHs (Harvey *et al.*, 1991)

On the other hand, PAHs are also quite involatile, and have relatively low vapor pressure and resistance to chemical reactions. As a consequence PAHs are persistent in the environment and demonstrate a tendency to accumulate in biota, soils, sediments, and are also highly dispersed by the atmosphere. Furthermore, PAHs become more resistant to biotic and abiotic degradation as the number of benzene rings increase (Bossert & Bartha, 1986). In addition, PAHs are generally not biomarkers as inferred by various authors (Bernstein et al., 1999; Ehrenfreund, 1999; Blake and Jenniskens, 2001), because they can form from abiological carbonaceous matter by free radical buildup (Simoneit and Fetzer, 1996) or by aromatization of biomarker natural products (Simoneit, 1998). The abiogenic PAHs occur primarily as the parent compounds without alkyl or alicyclic substituents as is the case for the aromatic biomarkers. Thus, phenanthrene is not a biomarker, but 1-methyl-7-phenanthrene (i.e., retene) is, because the latter is generated from numerous natural product precursors by dehydrogenation.

The 1952 Nobel Prize winners in chemistry, Archer John Porter Martin and Richard Laurence Millington Synge, had established the partition coefficient principle, which can be employed in paper chromatography, thin layer chromatography, gas phase and liquid-liquid applications. This technique is widely recognized as Hydrophilic Interaction Chromatography (HILIC) in HPLC or High-Performance Liquid Chromatography (sometimes referred to as High-Pressure Liquid Chromatography). HPLC characteristically exploits various kinds of stationary phases, a pump that drives the mobile phase(s) and analyte through the column, and a detector to afford a distinctive retention time for the analyte. Additional information related to the analyte can be achieved by using specific types of detectors (e.g. UV/Visible/Fluorescence Spectroscopic Detectors). It is also worth to mention that the analyte retention time is a function of its affinity with the stationary phase, the ratio/composition of solvents used, and the flow rate of the mobile phase. Although HPLC has the advantage of analyzing several organic compounds in a single chromatogram (e.g. analysis of PAHs), the sensitivity is not high enough to detect target compounds in part per billion (ppb) levels. To overcome the instrumental detection limit, Rolland Gohlke and Fred McLafferty had developed the application of a mass spectrometer as the detector in gas chromatography, which later on called as Gas Chromatography Mass Spectrometry (GC-MS) (Gohlke, 1959; Gohlke and Fred, 1993). In 1964, Robert E. Finnigan had extended the potential of GC-MS by developing a computer controlled quadrupole mass spectrometer under the collaboration with Electronic Associates, Inc. (EAI), a leading U.S. supplier of analog computers. By the 2000s computerized GC/MS instruments using quadrupole technology had become both essential to chemical research and one of the foremost instruments used for organic analysis. Today computerized GC/MS instruments are widely used in environmental monitoring of water, air, and soil; in the regulation of agriculture and food safety; and in the discovery and production of medicine.

Recently, GC/MS featuring ion trap systems, based on a selected parent ion and a whole mass spectrum of its daughter ions, offering the broadest selection of innovative features to expand the working mass range of typical GC/MS system to 900 amu with higher performances and specificities (Jie and Kai-Xiong, 2007). This technique minimizes the ratio between the numbers of ions from matrix background to those of target compound and thus enhances the analytical sensitivity of GC/MS system. Identifications of PAHs associated with atmospheric particulate matter, coastal and marine sediments are extremely elaborate since the congeners are generally present at trace levels in a complex mixture. Application of GC/MS featuring ion trap systems with low detection limits has allowed significant progress in characterization of organic component of particulate matter distributed in all environmental compartments. However, there are still some difficulties associated with the identification process using GC/MS featuring ion trap systems.

It is well known that the mass number of positively charged ions travelling through the quadrupole filter relies on the voltage applied to the quadrupole filter. As a consequence, the detection of a predetermined mass range can be conducted by altering (i.e. scanning) the applied voltage in a particular scope. Based on this theory, it is evident that the faster the scanning speed is, the shorter the detection time of a scanning cycle is, and the greater the number of scanning cycles within a predetermined time period can be achieved. This implies that a higher time resolution can be acquired by enhancing the scanning speed. However, the relatively fast scanning speed can lead to a major sensitivity drop due to the modification in the applied voltage while the positively charged ion is passing through the quadrupole. Assuming that a positively charged ion takes the time t1 to travel though the length L of the quadrupole filter, the kinetic energy of the ion plays an important role on the time t1 when the ion passes through the quadrupole filter. Since the voltage applied to the quadrupole while the ion passes through is positively correlated with the voltage scanning speed, the extremely high scanning speed deteriorates the instrument sensitivity as the scanning speed is enhanced.

GC-MS-QP 2010 Ultra was developed by Shimadzu engineers to resolve this problem. Advanced Scanning Speed Protocol technology (ASSP technology) was invented by using the electrical field to control the kinetic energy of the positively charged ions generated from the ion source while they are passing through the quadrupole filter. ASSP technology was specially designed for adjusting the kinetic energy of positively charged ions travelling through the quadrupole filter. The shorter the travelling time is, the alteration of voltage applied to rod electrodes in the course of the passage through the quadrupole filters is smaller and thus a large number of positively charged ions can enter to the quadrupole filter. This finally leads to an improvement of instrumental sensitivity. In addition, FASST (Fast Automated Scan/SIM Type) was deliberately designed for collecting Scan data and SIM data in a single measurement. This data acquisition technique can be employed in both gathering qualitative spectral data (Scan) and analyzing quantitative data (SIM). It is also worth to note that ASSP has amended this procedure by permitting the SIM dwell times to be reduced by as much as five times without losing sensitivity, letting the user to observe more SIM channels.

Without any doubt, ASSP coupled with FASST is an innovative data acquiring technique that integrates the qualitative assessment (i.e. Scanning of spectral data) with the quantitative analysis (SIM). Despite its high speed scanning performance, little is known about the factors governing relative response factors (RRFs) detected by GC-MS-QP 2010 Ultra (Shimadzu). Since the data quality is greatly reliant on the values of RRFs, it is therefore crucial for the analyzer to know the factors affecting sensitivity and stability of RRFs. An enormous number of papers have appeared on the topic of GC/MS analysis of PAHs using single quadrupole and quadrupole ion trap, most of which have not advanced knowledge to investigate the factors influence the sensitivity of RRFs determined by ASSP and/or FASST data acquiring techniques. Although the principles of the instrumental operation and data quality control are well established and straightforward, there are some uncertainties persisted and need to be clarified. For instance, there are several factors that can significantly affect the values of RRF detected by GC-MS ion trap such as "manifold temperature drop", "glass injection port linear contamination" and "column degradation" (Pongpiachan et al., 2009). However, it remained unclear to what extent other parameters such as ion source temperature (I.S.T), interface temperature (I.T) and scanning speed (S.S) could affect sensitivity of RRFs detected by using ASSP and FASST data acquiring techniques. Hence, the objectives of this study are

1. To statistically determine the effect of "ion source temperature" on fluctuations of RRFs of PAHs
2. To quantitatively assess the influence of "interface temperature" on alterations of RRFs of PAHs
3. To investigate the "scanning speed" effect on variations of RRFs of PAHs.

2. Materials and methods

2.1 Materials and reagents

All solvents are HPLC grade, purchased from Fisher Scientific. A cocktail of 15 PAHs Norwegian Standard (NS 9815: S-4008-100-T) (phenanthrene (Phe), anthracene (An), fluoranthene (Fluo), pyrene (Pyr), 11h-benzo[a]fluorene (11H-B[a]F), 11h-benzo[b]fluorene (11H-B[b]F), benz[a]anthracene (B[a]A), chrysene (Chry), benzo[b]fluoranthene (B[b]F), benzo[k]fluoranthene (B[k]F), benzo[a]pyrene (B[a]P), benzo[e]pyrene (B[e]P), indeno[1,2,3-cd]pyrene (Ind), dibenz[a,h]anthracene (D[a,h]A), benzo[g,h,i]perylene (B[g,h,i]P); each 100 µg mL-1 in toluene: unit: 1×1 mL) and a mix of recovery Internal Standard PAHs (d12-perylene (d12-Per), d10-fluorene (d10-Fl); each 100 µg mL-1 in xylene: unit: 1×1 mL) were supplied by Chiron AS (Stiklestadveine 1, N-7041 Trondheim, Norway). Standard stock solutions of 4 µg mL-1 of deuterated PAHs (used as internal standard) and 100µg mL-1 of native PAHs were prepared in nonane. Working solutions were obtained by appropriate dilution in n-cyclohexane. All solutions were stored in amber colored vials at -20 °C.

2.2 Analytical apparatus

A mass spectrometer is an instrument that separates ions according to the mass-to-charge ratio (m/z) and measures their relative abundance. The instrument is calibrated against ions of known m/z. All mass spectrometers operate by separating gas phase ions in a low-pressure environment by the interaction of magnetic or electrical fields on the charged particles. In this study, the analyses were performed using a Shimadzu GCMS-QP2010 Ultra system comprising a high-speed performance system with ASSP function (i.e. achieving maximum scan speed of 20,000 u sec-1) and an ultra-fast data acquisition speed for comprehensive two-dimensional gas chromatography (GC × GC). The rod bias voltage is automatically optimized during ultra high-speed data acquisition, thereby minimizing the drop in sensitivity that would otherwise occur above 10,000 u sec-1. The GCMS-QP2010 Ultra achieves a level of sensitivity better than five times that of conventional instruments, and is particularly effective for scan measurement in applications related to fast-GC/MS and comprehensive GC/MS (Patent: US6610979).

The target compounds were separated on a 60 m length × 0.25 mm i.d. capillary column coated with a 0.25 µm film thickness (phase composition: cross-linked/surface bonded 5% phenyl, 95% methylpolysiloxane. Specified in EPA methods 207, 508, 515, 515.2, 524.2, 525, 548.1, 680, 1625, 1653, 8081, 8141, 8270 and 8280) stationary phase (Agilent JW Scientific DB-5 GC columns). Helium (99.999%) was employed as carrier gas at a constant column flow of 1.0 mL min-1 and a pressure pulse of 25 psi with a duration of 0.50 min. All injections (1 µL) were performed through a universal injector in the splitless mode and the standards were introduced using a 10 µL Hamilton syringe. The GC oven temperature was programmed as follows: 1 min at at 40 °C, heated at 8 °C min-1 to 300 °C and held for 45 min. By employing these chromatographic conditions it was possible to qualitatively distinguish between Ph/An, B[a]A/Chry and 11H-B[a]F/11H-B[b]F, three pairs of isomers that are commonly co-eluted by gas chromatographic systems.

2.3 Compound identification and quantification

In order to quantify PAHs in environmental samples, all the detected compounds are normally identified by comparing the retention time and mass spectra of the authentic standards. Appropriately selected quantification ions can be beneficial to distinguish a particular mass spectrum of an individual compound from the co-eluted complex. In general, the most abundant ion serves as the quantification ion, which is the case for PAHs in this study. The molecular markers are identified by comparing first the retention times with authentic standards within a range of ± 0.2 min, secondly the quantification ions.

Quantification of the compounds is based upon the Internal Standard (IS) method. One of the fundamental requirements of using an IS is that it displays similar physiochemical properties or the same type of substitution as the analytes because be similar to each other. A relative response factor (RRF) for each native analyte was first determined. This is used for quantification, as the relative response between the internal standard (IS) and the native analyte should remain constant. It is a convenient method because recovery losses of the compound during extraction and analysis are assumed to match those of the IS. It is calculated employing the appropriate IS using the following equation: The calculation of relative response factor (RRF) is described as follows;

$$F = \frac{A_{nat}}{A_{is}} \times \frac{C_{is}}{C_{nat}} \quad (1)$$

Where A_{nat} = Peak area of the native compound in the standard; C_{nat} = Concentration of the native compound in the standard; A_{is} = Peak area of internal standard; C_{is} = Concentration of the internal standard. The RRF_{STD} used for quantifying samples are the mean of those calculated for the two quantification standards run on the same day. Concentration (C) of analytes in sample extracts is calculated using the following formula:

$$C = \frac{A_{nat}}{A_{is}} \times \frac{1}{RRF_{STD}} \times \frac{W_s}{W_{is}} \quad (2)$$

Where W_{is} = weight of IS added to the sample, W_s = weight or volume of the sample analyzed. A recovery determination standard (RDS) was used for the calculation of both internal standards (IS) and the sampling efficiency standard (SES) of recoveries during sample preparation and extraction/purification. A known amount of RDS was added at the final stage prior to GC/MS analysis and was assumed to suffer zero loss.

$$\left[\left(\frac{A_{is}}{A_{RDS}}\right)_S \times \left(\frac{A_{RDS}}{A_{is}}\right)_{STD}\right] \times \left[\left(\frac{C_{is}}{C_{RDS}}\right)_{STD} \times \left(\frac{C_{RDS}}{C_{is}}\right)_S\right] \times 100\% \quad (3)$$

Where A_{RDS} = Peak area of recovery determination standard; C_{RDS}: = Concentration of the recovery determination standard. Recoveries of IS were used as an indication of the anlyte losses during extraction, pre-concentration, cleanup/fractionation and blow down stages. The calculation of the sampling efficiency by using the sampling efficiency standard (SES) is described as follows:

$$\left[\left(\frac{A_{is}}{A_{SES}}\right)_{STD} \times \left(\frac{A_{SES}}{A_{is}}\right)_S\right] \times \left[\left(\frac{C_{is}}{C_{SES}}\right)_S \times \left(\frac{C_{SES}}{C_{is}}\right)_{STD}\right] \times 100\% \quad (4)$$

Where A_{SES} = Peak area of the sampling efficiency standard, C_{SES} = Concentration of the sampling efficiency standard. Recoveries of SES were used as an indication of analyte losses during sampling as opposed to the analysis. In this study, the IS d10-Fl and d12-Per were employed to calculate the values of RRFs of Group 1 PAHs (i.e. Phe, An, Fluo, Pyr, 11H-B[a]F, 11H-B[b]F, B[a]A, Chry) and Group 2 PAHs (i.e. B[b]F, B[k]F, Benzo[a]pyrene, B[e]P, Ind, D[a,h]A, B[g,h,i]P) respectively. In addition, the factors governing variations of RRFSTD of 15 PAHs Norwegian Standard (NS 9815: S-4008-100-T) as illustrated in Equation 1 had been carefully investigated. The results and details will be discussed in a later part of this chapter.

2.4 Statistical analysis

In this chapter, Pearson correlation analysis, multiple linear regression analysis (MLRA), hierarchical cluster analysis (HCA) and principal component analysis (PCA) were conducted by using SPSS version 13.

3 Results and discussion

Ion source temperature (I. S. T), interface temperature (I. T) and scanning speed (S. S) were set at five different levels (i.e. 200 °C, 225 °C, 250 °C, 275 °C, 300 °C), four distinctive stages (i.e. 250 °C, 275 °C, 300 °C, 325 °C) and five altered points (500 u sec-1, 1,000 u sec-1, 5,000 u sec-1, 10,000 u sec-1, 20,000 u sec-1). Therefore, there are 100 combinations of I.S.T, I.T and S.S (i.e. 4×5×5 = 100) included in the statistical analysis.

3.1 Pearson correlation analysis

To find the relationship between peak area of PAHs and GC-MS tuning parameters (i.e. I.S.T, I.T and S.S), the Pearson correlation coefficients were computed and displayed in Table 3. Some comparatively strong negative correlations were observed between two continuous variables, namely S.S. and high molecular weight (HMW) PAHs. For instance, R-values of S.S. vs Benzo[a]pyrene, S.S. vs Ind, S.S. vs D[a,h]A were -0.77, -0.76, -0.75 respectively. As discussed earlier in Section 1, the relatively fast scanning speed can dramatically cause a significant drop of instrumental sensitivity. Therefore, ASSP and FASST were developed for resolving this technical problem. By carefully controlling the kinetic energy of positively charged ions travelling inside the mass spectrometry, one can reduce its passing time and thus allowing more ions to enter the quadrupole filter. However, if the kinetic energy is not high enough to shorten the passing time of positively charged ions inside the quadrupole, a major drop of instrumental sensitivity can be detected as previously mentioned. Some positive correlations between S.S. vs Phe (R = 0.61), An (R = 0.65), deut-Fl (R = 0.63) proved that ASSP and FASST seem to work fairly well on LMW PAHs (see Table 3). This can be explained through a competition between the kinetic energy of positively charged ions and the voltage scanning speed. It appears that ASSP and FASST successfully resolved the sensitivity drop problems for LMW PAHs. In addition, neither I.S.T. nor I.T. play a significant role on the variation of PAHs' sensitivities.

3.2 Cluster analysis

Cluster analysis (CA), also called segmentation analysis or taxonomy analysis, seeks to identify homogeneous subgroups of cases in a population. That is, cluster analysis seeks to identify a set of groups, which both minimize within-group variation and maximize between-group variation. In this study, CA was conducted using SPSS 13.0 for Windows. CA techniques may be hierarchical (i.e. the resultant classification has an increasing number of nested classes) or non-hierarchical (i.e. k-means clustering). Hierarchical clustering allows users to select a definition of distance, then select a linking method of forming clusters, then determine how many clusters best suit the data. Hierarchical clustering methods do not require pre-set knowledge of the number of groups.

	I.S.T.*	I.T.**	S.S.***	Ph	An	deut-Fl	Fluo	Pyr	11H-B[a]F	11H-B[b]F	B[a]A	Chry	B[b]F	B[k]F	B[e]P	B[a]P	Ind	D[a,h]A	B[g,h]P
I.S.T.	1																		
I.T.	-1.4E-17	1																	
S.S.	2.4E-17	1.76E-17	1																
Ph	0.289	0.160	**0.610**	1															
An	0.284	0.113	**0.647**	**0.984**	1														
deut-Fl	0.189	0.117	**0.632**	**0.979**	**0.991**	1													
Fluo	0.457	0.208	0.356	**0.914**	**0.901**	**0.877**	1												
Pyr	0.454	0.198	0.338	**0.904**	**0.889**	**0.866**	**0.989**	1											
11H-B[a]F	0.485	0.243	-0.188	0.509	0.465	0.435	**0.783**	**0.798**	1										
11H-B[b]F	0.456	0.251	-0.202	0.477	0.430	0.396	**0.756**	**0.769**	**0.975**	1									
B[a]A	0.315	0.240	-0.687	-0.121	-0.179	-0.207	0.239	0.262	**0.754**	**0.777**	1								
Chry	0.397	0.262	-0.574	0.040	-0.012	-0.047	0.393	0.415	**0.846**	**0.856**	**0.965**	1							
B[b]F	0.302	0.204	**-0.777**	-0.333	-0.388	-0.419	0.025	0.046	0.598	**0.630**	**0.955**	**0.898**	1						
B[k]F	0.242	0.220	**-0.746**	-0.341	-0.393	-0.426	0.006	0.029	0.574	**0.601**	**0.941**	**0.895**	**0.951**	1					
B[e]P	0.268	0.219	**-0.779**	-0.346	-0.401	-0.432	0.011	0.034	0.592	**0.621**	**0.961**	**0.906**	**0.986**	**0.977**	1				
B[a]P	0.306	0.194	**-0.766**	-0.324	-0.378	-0.413	0.037	0.060	**0.610**	**0.634**	**0.964**	**0.914**	**0.987**	**0.974**	**0.993**	1			
Ind	0.313	0.200	**-0.756**	-0.309	-0.365	-0.400	0.052	0.070	**0.618**	**0.641**	**0.953**	**0.911**	**0.974**	**0.959**	**0.978**	**0.980**	1		
D[a,h]A	0.391	0.183	**-0.753**	-0.319	-0.372	-0.413	0.045	0.065	**0.612**	**0.632**	**0.948**	**0.906**	**0.982**	**0.962**	**0.983**	**0.988**	**0.981**	1	
B[g,h]P	-0.022	0.185	**-0.621**	-0.195	-0.250	-0.245	0.077	0.085	0.517	0.544	**0.763**	**0.724**	**0.733**	**0.730**	**0.748**	**0.738**	**0.825**	**0.713**	1
deut-Per	0.257	0.181	**-0.795**	-0.400	-0.452	-0.482	-0.044	-0.022	0.552	0.576	**0.943**	**0.883**	**0.981**	**0.976**	**0.990**	**0.989**	**0.978**	**0.983**	0.740

Table 3. The Pearson correlation coefficients of PAH's peak area (PA)

There are three general approaches to clustering groups of data, namely "Hierarchical Cluster Analysis", "K-means Cluster Analysis" and "Two-Step Cluster Analysis". In both K-means clustering and two-step clustering, researchers have to specify the number of clusters in advance then calculate how to assign cases to the K clusters. Furthermore, these two clustering techniques require a very large scale of data set (e.g. n > 1,000). On the contrary, Hierarchical Cluster Analysis (HCA) is appropriate for smaller samples (e.g. n < 200) and can be carried out without any data pre-treatment. It is also important to note that there are two types of Hierarchical Clustering namely Agglomerative Hierarchical Clustering and Divisive Hierarchical Clustering.

In agglomerative hierarchical clustering every case is initially considered as a cluster then the two cases with the lowest distance (i.e. highest similarity) are combined into a cluster. The case with the lowest distance to either of the first two is considered next. If that third case is closer to a fourth case than it is to either of the first two, the third and fourth cases become the second two-case cluster, if not, the third case is added to the first cluster. The process is repeated, adding cases to existing clusters, creating new clusters, or combining clusters to get to the desired final number of clusters. In contrast, the divisive clustering works in the opposite direction, starting with all cases in one large cluster. Since the

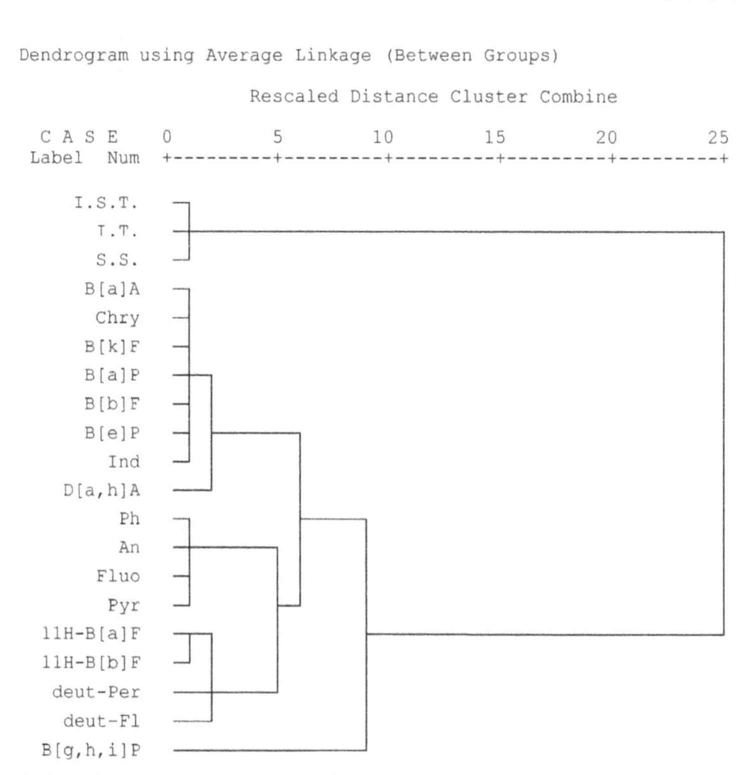

Fig. 1. Dendrogram using average linkage between group of PAH's peak area, I.S.T, I.T and S.S.

objective of this section is to clusterize the wide range of GC-MS tuning parameters coupled with RRFs of PAHs with relatively small sample numbers, the agglomerative hierarchical cluster analysis is probably the most suitable for this purpose. HCA was carried out using the multivariate data analysis software "SPSS version 13". To obtain more information on the affinities of GC-MS tuning parameters and PAHs' peak area, HCA had been conducted on the 20 variables (i.e. 17 PAHs and 3 GC-MS tuning parameters). The results demonstrated in the dendrogram (Fig. 1) distinguished the 20 individual parameters into two major clusters. The first cluster consists mainly of I.S.T, I.T and S.S, which can be considered as a domination of GC-MS tuning parameters.

The second major cluster can be subdivided into three sub-clusters. The first sub-cluster consists of B[a]A, Chry, B[k]F, Benzo[a]pyrene, B[b]F, B[e]P, Ind, D[a,h]A, which were all composed of four to five benzene rings. Therefore, this sub-cluster represents the typical marks of HMW PAHs. The second sub-cluster contains LMW PAHs (Ph, An, Fluo, Pyr). The third sub-cluster composed of 11H-B[a]F, 11H-B[b]F, deut-Per and deut-Fl. It is interesting to note that these congeners have different chemical structures from those of other PAHs. Therefore, the third cluster can be considered as an indicative of mixing of unique structures of PAHs.

Since dendrogram provides a visual accounting of how closely connected one parameter is to another, the more properties of two parameters have in common, the closer they are associated. Figure 1 demonstrates that, for this dendrogram of four clusters, there is a fairly close-knit group in the first major cluster (i.e. I.S.T, I.T and S.S) and the first sub-cluster of second major cluster (i.e. B[a]A, Chry, B[k]F, Benzo[a]pyrene, B[b]F, B[e]P, Ind, D[a,h]A). B[g,h,i]P is the outsider in this dendrogram, while the second sub-cluster of second major cluster (i.e. Ph, An, Fluo, Pyr) takes an intermediary position. Similar to previous results obtained by using Pearson correlation analysis, S.S. is still the most influential GC-MS tuning parameters governing the instrumental sensitivity of PAHs. Apart from B[g,h,i]P, the four-to-five aromatic ring PAHs are closely associated with the fluctuations of S.S. emphasizing that one should carefully select the suitable scanning speed in order to minimize the sacrifice of instrumental sensitivity.

3.3 Multiple linear regression analysis

Table 4 displays the standardized coefficients of multiple linear regression analysis (β-value) of PAH's relative response factors (RRFs). It is remarkable to note that the moderately strong negative β-values (i.e. -0.78 ~ -0.83) and fairly strong positive β-values (i.e. 0.68 ~ 0.76) were observed with S.S as displayed in Fig 4. These results are consistent with previous findings computed by using Pearson correlation analysis and hierarchical cluster analysis. On the contrary, the comparatively weak β-values (i.e. -0.78 ~ -0.83) were detected with I.T. as displayed in Fig 3. It is well known that too high interface temperature can considerably degrade the chemical structure of target compound and hence reducing the instrumental sensitivity. On the other hand, if the interface temperature was too low, the cold condensation will occur and conclusively undermining the instrumental sensitivity of PAHs. Interestingly, there are no obvious signs of either "thermal decomposition" or "cold condensation" which can appreciably degrade the instrumental sensitivity of PAHs. In addition, I.S.T. show a fairly strong positive β-values with Ph (0.38), An (0.70), Ind (0.40) and D[a,h]A (0.72) as illustrated in Fig. 4.

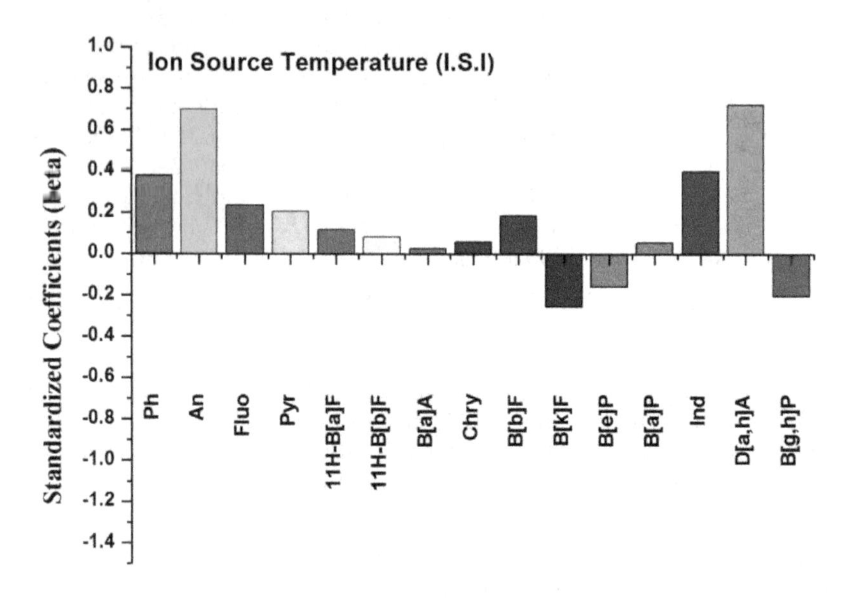

Fig. 2. The standardized coefficients of multiple linear regression analysis (MLRA) of PAH's relative response factors (RRFs) (Ion Source Temperature)

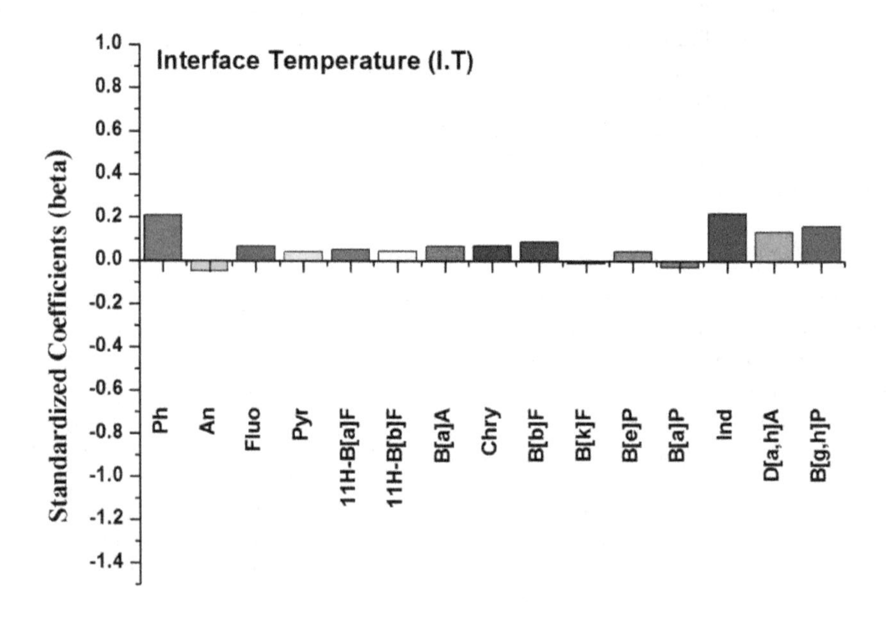

Fig. 3. The standardized coefficients of multiple linear regression analysis (MLRA) of PAH's relative response factors (RRFs) (Interface Temperature)

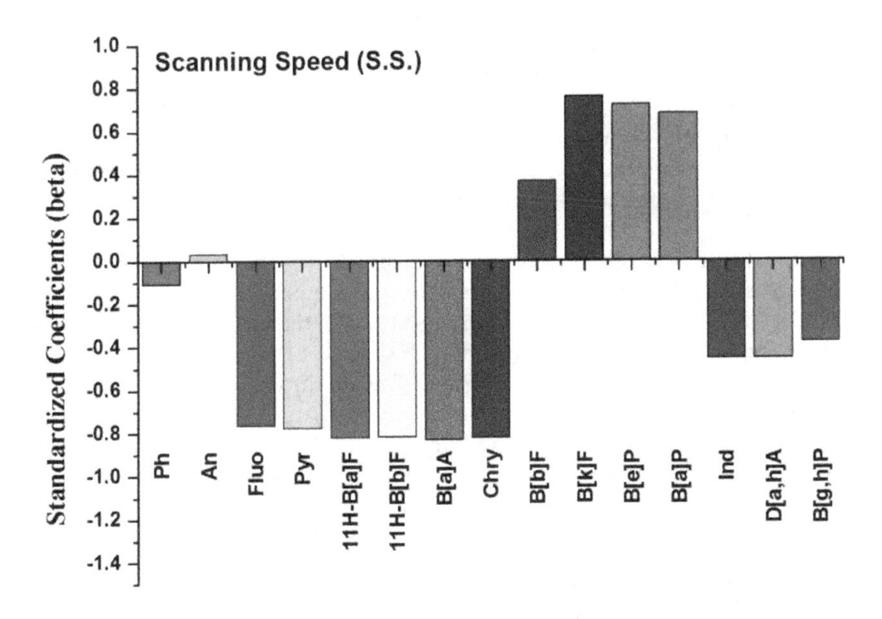

Fig. 4. The standardized coefficients of multiple linear regression analysis (MLRA) of PAH's relative response factors (RRFs) (Scanning Speed)

	Ph	An	Fluo	Pyr	11H-B[a]F	11H-B[b]F	B[a]A	Chry	B[b]F	B[k]F	B[e]P	B[a]P	Ind	D[a,h]A	B[g,h]P
I.S.T.*	0.379	**0.698**	0.237	0.204	0.117	0.080	0.024	0.057	0.184	-0.255	-0.158	0.055	0.401	**0.721**	-0.203
I.T.**	0.209	-0.047	0.067	0.040	0.053	0.044	0.066	0.069	0.086	-0.013	0.043	-0.031	0.219	0.134	0.160
S.S.***	-0.110	0.030	**-0.766**	**-0.775**	**-0.824**	**-0.818**	**-0.832**	**-0.822**	0.367	**0.760**	**0.721**	**0.682**	-0.457	-0.458	-0.381

*I.S.T: Ion Source Temperature, **I.T: Interface Temperature, ***S.S: Scanning Speed

Table 4. The standardized coefficients of multiple linear regression analysis (MLRA) of PAH's relative response factors (RRFs)

3.4 Principal Component Analysis

PCA as the multivariate analytical tool is employed to reduce a set of original variables (i.e. GC-MS tuning parameters and PAHs' RRFs) and to extract a small number of latent factors (principal components, PCs) for analyzing relationships among the observed variables. Data submitted for the analysis were arranged in a matrix, where each column corresponds to one parameter and each row represents the number of data. Data matrixes were evaluated through PCA allowing the summarized data to be further analyzed and plotted. Table 5 displays the principal component patterns for Varimax rotated components of I.S.T., I.T., S.S. and PAHs'RRFs. In order to enable further interpretation of potential correlations between GC-MS tuning parameters and PAHs' RRFs, a PCA model with five significant PCs, each representing 54.08 %, 14.91 %, 9.42 %, 5.61 % and 4.72 % of the variance, thus accounting for 88.74 % of the total variation in the data, was calculated. The first component (PC1) shows considerable strong negative correlation coefficients on five aromatic ring PAHs (i.e. B[k]F, B[e]P, Benzo[a]pyrene) and strong positive correlation coefficients on three-to-four aromatic ring PAHs (i.e. Fluo, Pyr, 11H-B[a]F, B[a]A, Chry). These results confirm that GC-MS tuning parameters do significantly affect these two classes of PAHs in a totally different way. The strong negative correlation coefficient (R = -0.87) detected in S.S. reflected the fact that higher scanning velocity reduced numbers of positively charged ions of HMW PAHs and thus decreased the instrumental sensitivity of five aromatic compounds. The second component (PC2) has higher loadings for I.S.T. (R = 0.89), Ph (R = 0.61), An (R = 0.87) and D[a,h]A (R = 0.65), which could be due to the fact that higher ion source temperature enhanced the signal of three ring aromatic compound (i.e. Ph, An) and five ring aromatic PAHs (i.e. D[a,h]A). The third component (PC3), the fourth component (PC4) and the fifth component (PC5) shows high loading of Ind (R = 0.81), B[b]F (R = 0.77) and I.T. (R = 0.91) respectively. Since I.T. showed the highest R-value in PC5, it appears reasonable to conclude that interface temperature plays a minor role in fluctuations of PAHs' signals detected by the GCMS-QP2010 Ultra. In addition, some HMW PAHs such as Ind and B[b]F appears to insensitive to the variations of GC-MS tuning parameters.

	Principal Component				
	PC1	PC2	PC3	PC4	PC5
I.S.T.	0.034	**0.886**	0.017	0.279	-0.107
I.T.	0.024	-0.036	0.096	0.105	**0.906**
S.S.	**-0.866**	0.096	-0.269	-0.142	0.034
Ph	0.093	**0.610**	0.207	-0.177	0.448
An	0.148	**0.869**	-0.146	-0.136	-0.043
Fluo	**0.926**	0.296	0.072	-0.084	0.091
Pyr	**0.936**	0.259	0.046	-0.083	0.073
11H-B[a]F	**0.965**	0.159	0.099	-0.084	0.070
11H-B[b]F	**0.962**	0.134	0.080	-0.077	0.081
B[a]A	**0.969**	0.070	0.043	-0.095	0.104
Chry	**0.965**	0.104	0.050	-0.113	0.099
B[b]F	-0.499	0.080	-0.028	**0.772**	0.116
B[k]F	**-0.840**	-0.151	-0.242	-0.171	0.109
B[e]P	**-0.844**	-0.116	-0.192	0.175	0.156
B[a]P	**-0.821**	0.104	-0.160	0.290	0.071
Ind	0.316	0.396	**0.810**	0.186	0.136
D[a,h]A	0.452	**0.646**	0.249	0.456	0.061
B[g,h]P	0.166	-0.185	0.937	-0.115	0.084
% of Variance	54.083	14.913	9.420	5.613	4.724

Table 5. Principal component analysis of PAH's relative response factors (RRFs)

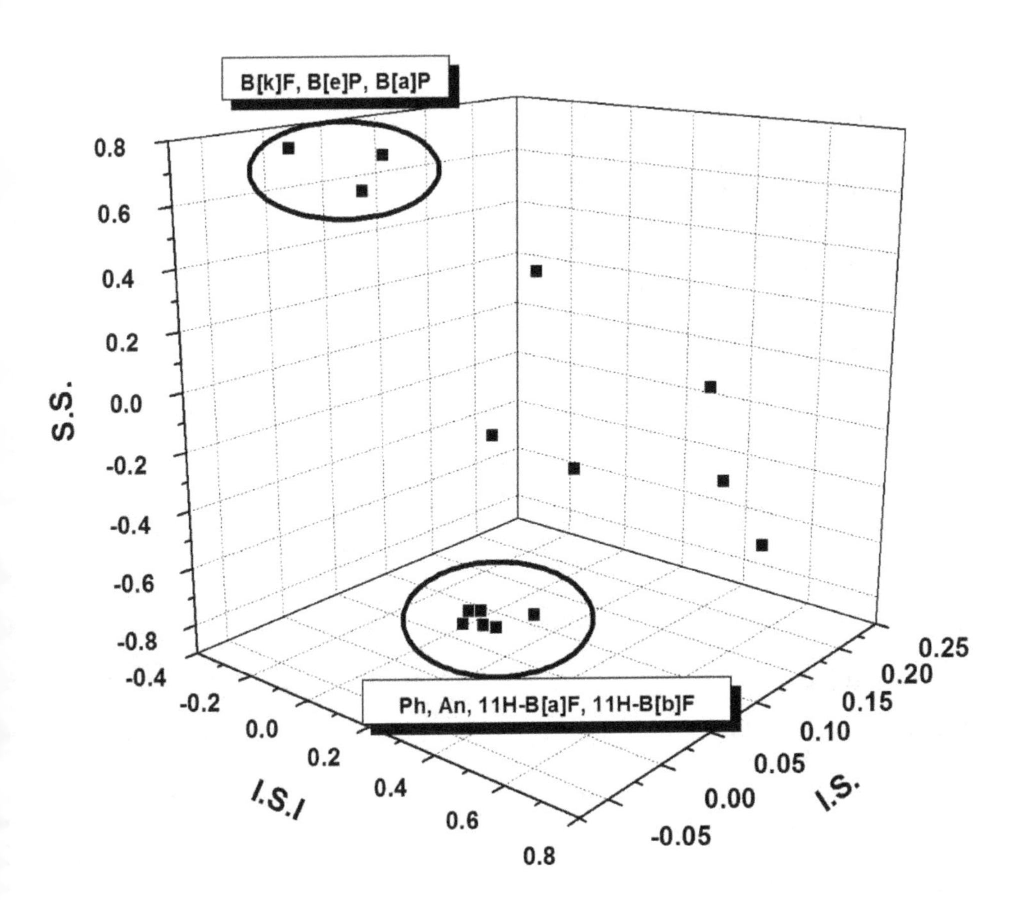

Fig. 5. Three dimensional plots of standardized coefficients of I.S.I, I.S. and S.S.

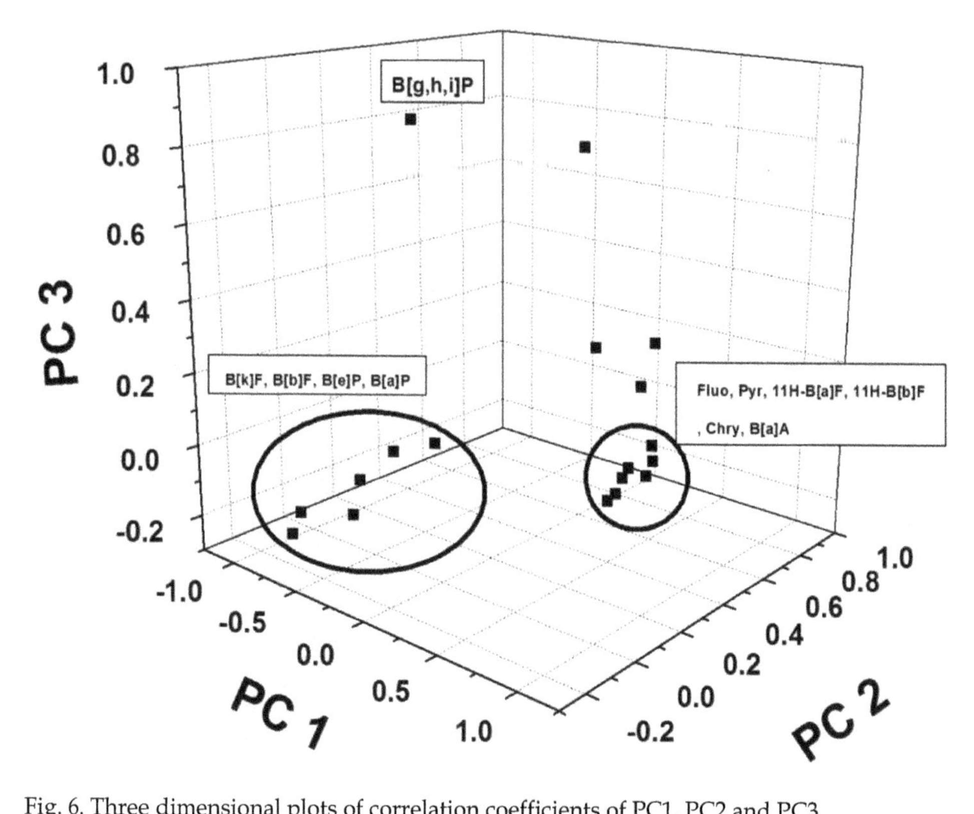

Fig. 6. Three dimensional plots of correlation coefficients of PC1, PC2 and PC3

4. Conclusions

This study is the first, to our best knowledge, to be presented for the quantitative approach to investigate the effects of "ion source temperature", "interface temperature" and "scanning speed" on variations of RRFs of 15 PAHs determined by Shimadzu GCMS-QP2010 Ultra. In fact, the impact of "ion source temperature" and "interface temperature" affect only a few highly volatile PAHs, whereas the effects of "scanning speed" significantly cause the variations of instrumental sensitivities of PAHs. Despite of various factors deteriorating sensitivity of RRFs, the statistical analysis reflect the considerable contribution of scanning speed on relative sensitivities of PAHs in both positive and negative directions. Since the reliability of analytical results are highly dependent on the "sensitivity" of RRFs, it is of great importance for the analyzer to ensure the correct selection of "scanning speed" suitable for the target compounds.

5. Acknowledgement

The authors would like to thank Bara Scientific Co., Ltd and NIDA Center for Research and Development on Natural Disaster Managment, School of Social and Environmental

Development, National Institute of Development Administration (NIDA) for the continued financial support of this work.

6. References

Bernstein, M.P., Sandford, S.A. and Allamandola, L.J. (1999). Life's far-flung raw materials. *Sci. Amer.* 281 1, pp. 42–49.

Blake, D.F. and Jenniskens, P. (2001). The ice of life. *Sci. Amer.* 285 (2): 44–50.

Bossert, I.P. and Bartha, R. (1986). *Bull. Environ. Contam. Toxicol.*, 37: 490-495.

Ehrenfreund, P (1999) Molecules on a space odyssey. *Science* 283: 1123–1124.

EPA Appendix A to 40 CFR, Part 423–126 Priority Pollutants Available from: http://www.epa.gov/region01/npdes/permits/generic/prioritypollutants.pdf, (2003).

Gohlke, R. S. (1959). Time-of-Flight Mass Spectrometry and Gas-Liquid Partition Chromatography. *Analytical Chemistry* 31 (4): 535. doi:10.1021/ac50164a024.

Gohlke, R and McLafferty, F. W. (1993). Early gas chromatography/mass spectrometry. *Journal of the American Society for Mass Spectrometry* 4 (5): 367. doi:10.1016/1044-0305(93)85001-E.

Harvey, G. R (1991) *Polycyclic aromatic hydrocarbons,* Cambridge University Press, Cambridge, USA.

International Agency for Research on Cancer (1983) IARC Monographs on the evaluation of the carcinogenic risk of chemicals to man, Vol.32: Polycyclic aromatic hydrocarbons, Part 1: Chemical, environmental and experimental data. IARC, Lyon, France.

International Agency for Research on Cancer (1984) IARC Monographs on the evaluation of the carcinogenic risk of chemicals to man, Vol.34: Polycyclic aromatic hydrocarbons, Part 3: Industrial exposures in aluminium production, coal gasification, coke production, and iron and steel founding. IARC, Lyon, France.

International Agency for Research on Cancer (1985) IARC Monographs on the evaluation of the carcinogenic risk of chemicals to man, Bitumens, coal-tars and derived products, shale-oils and soots. *Monnograph No 35,* IARC, Lyon, France.

International Agency for Research on Cancer (1987) IARC Monographs on the evaluation of the carcinogenic risk of chemicals to man, Supplement, IARC, Lyon, France.

Jie, F and Kai-Xiong, W. (2007). Multiresidual Analysis of Organochlorine Pesticides, Polychlorinated Biphenyls and Polycyclic Aromatic Hydrocarbons in Marine Shellfishes by Gas Chromatography-Ion Trap Mass Spectrometry. *Chinese J Anal Chem* 35 (11): 1607-1613.

Mackay, D., Callcot, D (1998) Partitioning and physical properties of PAHs. In: Neilson, A.H. (Ed.), The Handbook of Environmental Chemistry, vol. 3, Part J. PAHs and Related Compounds. Springer, Berlin, pp. 325–346.

Pongpiachan, S., Bualert, S., Sompongchaiyakul, P and Kositanont, C. (2009). Factors affecting sensitivity and stability of polycyclic aromatic hydrocarbons. *Journal of Analytical Letters* 42 (13): 2106-2130.

Simoneit, B.R.T. and Fetzer, J.C (1996) High molecular weight polycyclic aromatic hydrocarbons in hydrothermal petroleums from the Gulf of California and Northeast Pacific Ocean. *Org. Geochem* 24: 1065–1077.

Simoneit, B.R.T. (1998). Biomarker PAHs in the environment. In: Neilson, A. and Hutzinger, O., Editors, *The Handbook of Environmental Chemistry*, Springer Verlag, Berlin: 175–221.

Wild, S.R. and Jones, K.C. (1995) Polynuclear Aromatic Hydrocarbons in the United Kingdon Environment: A Preliminary Source Inventory and Budget. *Environ. Pollut.* 88: 91-108.

Part 2

Selected Biomedical Applications of Gas Chromatography Techniques

7

Analysis of Toxicants by Gas Chromatography

Sukesh Narayan Sinha[1,*] and V. K. Bhatnagar[2]
[1]*National Institute of Nutrition (ICMR), Jamia-Osmania, P.O-Hyderabad,*
[2]*Principal and Dean, Subharti Institute of Technology & Engineering, Swami Vivekanand Subharti University, Meerut, India*

1. Introduction

Chromatography is an analytical technique which has been used for isolation, purification and separation of organic and inorganic compounds including qualitative and quantitative estimation of compounds. Basically, there are two types of chromatography. One is liquid chromatography and other is Gas chromatography. The chromatography was discovered in 1906 and gas chromatography was developed by Tshett in 1950s. Generally, in GC the mobile phases are gases such as helium (He), or Nitrogen (N_2). GC depends upon temperature programming and boiling point of the compounds. The volatile organic compounds (VOCs), poly-aromatic hydrocarbon and pesticides have been analysed by gas chromatography-mass spectrometric technique. In this chapter, we present a review of existing analytical methodology for the environmental and biological monitoring of exposure to toxicants. In an effort to provide a more concise exploration of existing toxicants bio-monitoring methodology that is relevant today. In addition, as much of the toxicants analysis work involving effective dose measurements is its infancy (adduct measurements) or do not provide sensitive indicators of exposure, All these methods are in use but some form of chromatography, but the detection systems range from simple UV absorbance detection to sophisticated mass spectrometric analyses. These methods possess limits of detection (LODs) that span a wide range; some are suitable for only occupational or forensic applications while those with LODs near or lower than the low-µg/ l are useful for detecting incidental environmental exposures. In addition, these methods have been used to measure toxicants and/or their metabolites in a variety of matrices including urine, serum, breast milk, saliva, and post-partum meconium.

2. Gas chromatography equipped with different detectors for analysis of Arial pollutants

Generally, five types of detectors are coupled with GC (GC-ECD, GC-NPD, GC-FID, GC-PID and GC-MS). The analysis of compounds on GC depends upon nature of compound,

* Corresponding Author

physical and chemical properties of compound. Gas chromatography coupled with electron capture detector (GC-ECD) is used for qualitative as well as quantitative estimation of organochlorine compounds especially organochlorine pesticides. Gas chromatography coupled with nitrogen phosphorous detector (GC-NPD) is used for quantification of nitrogen and phosphorous containing compounds, especially organophosphate compounds. Gas chromatography coupled with flame ionization detector (GC-FID) is used for quantification of volatile organic compounds. Gas chromatography mass spectrometer (GC-MS) is used for quantitative as well as qualitative estimation of non-polar compounds. This technique is very useful for structural confirmation as well as molecular weight determination of compounds. Now, mass spectrometer is used for measurement of molecular weight as well as structural estimation and quantification of compounds. Various types of MS may be used for analysis of protein, peptide, mapping of oligonucleotides, pesticides and air pollutants. MS is also used in industrial as well as pharmaceutical application in drug discovery, poly-aromatic hydrocarbons, poly-chlorinated biphenyl, clinical screening, haemoglobin analysis and drug testing). GC-MS is now a well established routine technique where the MS is viewed as another detector and used for analysis of air samples. GC-MS (eg. High resolution GC-MS (Electron impact (EI), chemical ionization (CI) and negative chemical ionization (NCI) (Ballschmiter et al., 1992; Fischer et al., 1992) techniques were used for the determination of relevant environment parameters (eg. PO, SH_20, KOW) in the mixture of organic compound. This technique was also used to study the chemistry of incomplete combustion and high temperature pyrolisis. The emission from the technical burning process, including the organic emissions of automobile and their occurrence in environmental samples were studied for the groups of polychlorinated, dibenzofuran, biphenyls, diphenyl ethers, chlorophenols and microbial degradation of polychlorinated pesticides and arenes like the dibenzo-p-dioxins, dibenzofurans and naphthalines. A combined GC-MS technique is used in USA for the Qualitative and quantitative analysis of methyl terbutyl ether (MTBE) and benzene in gasoline (Star et al., 2000; Quach et al., 1998). The GC-MS experiments demonstrate the use of internal standards in the method to improve precision, standard deviation, and ion extraction/monitoring for the measurement of specific highly volatile organic compounds in air samples. The thermally desorbed compounds were analysed by GC/FID (flame ionization detector)/MS utilising a GC oven program ranging from 2^0C (to prevent column plugging by ice) to 160^0C and separation on a 30 m x 0.32 mm DB5-MS capillary column. VOCs are identified on the basis of retention time and mass spectral search, where as quantification is based on FID response which should be calibrated weekly with VOC standards (Hodgson et al., 1989; Reidel,1996). The FID detector is more sensitive than MS detector for analysis of VOC's. Structural characterization of PAC by combined GC-MS and GC-FTIR provides unambiguous identification of the isomers in iomex samples.

GC-MS was used to quantify triphenyl hydroquinol THP (Abdulrahman (2001) in environmental samples. Analysis of these data shows that there is no statistical difference between TPH values quantified with GC-MS and those derived from conventional TPH methods with GC/FID. Further, the GC-MS technique used in the study is readily adaptable to most environmental laboratories for analysis of volatile and semi-volatile compounds (EPA analysis ie 8260 and 8270 method respectively, EPA 845 (Robert, (1993). The implications of these results are 1) additional information (e.g. PAH) is often required when conducting a risk based assessment which can be derived from pre-existing site assessment data, thereby decreasing the cost and time required to obtain the additional samples and analytical information, and (2) unique TPH distributions can be critically evaluated with mass spectra in order to ascertain the nature and potential source of the compounds present in the TPH mixtures.

A gas chromatography–mass spectroscopic method in electron ionization (EI) mode with MS/MS ion preparation using helium at flow rate 1 ml min^{-1} as carrier gas on DB-5 capillary column (30 m × 0.25 mm i.d. film thickness 0.25 μm) has been developed for the determination of benzene in indoor air. The detection limit for benzene was 0.002 μg ml^{-1} with S/N: 4 (S: 66, N: 14). The benzene concentration for cooks during cooking time in indoor kitchen using dung fuel was 114.1 μgm^{-3} while it was 6.6 μgm^{-3} for open type kitchen. The benzene concentration was significantly higher ($p < 0.01$) in indoor kitchen with respect to open type kitchen using dung fuels. The wood fuel produces 36.5 μgm^{-3} of benzene in indoor kitchen. The concentration of benzene in indoor kitchen using wood fuel was significantly ($p < 0.01$) lower in comparison to dung fuel. This method may be helpful for environmental analytical chemist dealing with GC–MS in confirmation and quantification of benzene in environmental samples with health risk exposure assessment (Sinha et al., 2005).

The benzene and toluene was quantified using the GC-MS/ MS technique from indoor air. The geometrical mean (GM) of benzene exposure for cooks during cooking hours in an indoor kitchen using mixed fuel was 75.3 $\mu g/m^3$ (with partition) and 63.206 $\mu g/m^3$ (without partition), while the exposure was 1.7 $\mu g/m^3$ for open type. The benzene exposure was significantly higher ($p < 0.05$) in an indoor kitchen with respect to open type using mixed fuels. Concentration of benzene (114.1 $\mu g/m^3$) for cooks in an indoor kitchen with partition using dung fuel was significantly higher in comparison to open type kitchen. (Sinha et al., 2005). The toxic compounds are toluene, xylene, ethylbenzene, 2-4dimethyl heptanes ,cyclohexane propanol, 1-(methyl ethyl) cyclohexane, propyl benzene, 1-ethyl-3-methyl benzene, heptyl cyclohexane, 1-methyl-3propyl benzene, diethyl benzene, 4-ethyl octane, naphthalene and 3-methyl decane were present in thinner. All these compounds were confirmed using retention time parameter, molecular ion peak and other characteristic peak using National Institute of Science Technology (NIST) library search. The percentage composition of thinner has been also explained (Sinha and Zadi 2008). A GC-MS method is also very useful for measurement of main urinary metabolites of benzene, namely phenol, catechol,hydroquinone, 1,2,4-trihydroxybenzene, t,t-muconic acid and S-phenylmercapturic acid. Measurement of urinary benzene was performed via headspace solid phase microextraction of 0.5 mL of urine specimens followed by GC-MS. A number of methods including GC-FID and GC-MS are available for analysis of benzene from ambient environment but the more sensitive procedure of Pre-concentration on charcoal followed by GC-MS analysis is preferred. GC-MS technique being used for the qualitative and quantitative analysis for methyl tertiary butyl ether (MTBE) and benzene in gasoline. In this method, the internal standard was used to improve precision, standard deviation and ion extraction/ monitoring for the measurement of specific highly volatile organic compounds in air pollution samples (Carmichael et al., 1990). A carbotrap tube (2 mm) was used to determine volatile organic compounds in ambient air. Such compounds were desorbed and thermally analyzed with GC-MS With the use of GC-MS technique about 54 toxic hydrocarbons were quantified in the ambient air of Tehran. Polyurethane foam (PUF) cartridge samples were analyzed for dioxins and furans as per EPA test method TO-9 (Keen and Doug project No 46310001131). This gas chromatography high resolution mass spectrometric (GC-HRMS) method was also used for the analysis of poly chlorinated dibenzofurans in ambient air.

The gas chromatography electron ionization with single ion monitoring (GC-EI/SIM), gas chromatography-negative chemical ionization with single ion monitoring (GCNCI/SIM), and gas chromatography-negative chemical ionization with single reaction monitoring (GC-NCI/SRM) method was developed for the determination of pesticides in air sample extracts at concentrations <100-μgL^{-1}. In general, GC-NCI/SIM provided the lowest method detection limits (MDLs commonly 2.5–10 μgL^{-1}), while GC-NCI/SRM was used for confirmation of parathion-ethyl, tokuthion, carbofenothion. But, as per the reported method GC-EI/ SRM at concentration <100- μgL^{-1} was not suitable for most pesticides. GC-EI/SIM was more prone to interference issues than negative chemical ionization (NCI) methods, but of fixed good sensitivity (MDLs 1–10 μgL^{-1}) for pesticides with poor NCI response (organophosphates (Ops): sulfotep, phorate, aspon, ethion, and OCs: alachlor,aldrin, perthane, and DDE, DDD, DDT) (Martinez et al., 2004).

3. Chromatography equipped with different detector for analysis of pesticides in food samples

GC–MS/MS method was used for the analysis of organochlorine pesticides (OCs) residues (α-HCH, β-HCH, γ-HCH, δ-HCH, p,p-DDE, p,p-DDD, p,p-DDT, α-endosulfan, β-endosulfan, and endosulfan sulphate) in carbonated drinks (Miller and Miline 2008). Furthermore, a method was reported (Paya et al., 2007) for the analysis of pesticide residues using a quick, cheap, effective, rugged, and safe (QuEChERS) multi-residue method in combination with gas chromatography and tandem mass spectrometric detection (GC–MS/MS). A mixture of 38 pesticides was quantitatively recovered from spiked lemon, raisins, wheat and flour using GC–MS/MS, while 42 pesticides were recovered from oranges, red wine, red grapes, raisins and wheat flour using GC–MS/MS for determination (Paya et al., 2007). Several GC methods have also been developed for pesticide residual analysis in different food (Kumari and Kathpal 2009; Kumari et al., 2006; kumara et al., 2002) commodities (e.g., vegetables, fruits and other products of food).

A sensitive method for the quantification of 11 pesticides in sugar samples to the μgkg^{-1} level has been developed. These pesticides are often used in an agricultural context. A simple solvent extraction followed by selective analysis using a gas chromatography–mass spectrometric method was used. This method was accurate (> 99%) as it possesses limits of detection in the 0.1-ug kg^{-1} range, and the coefficients of variations are less than 15% at the low μgkg^{-1} end of the method's linear range. The percent recovery of all the pesticides at the lowest levels of detection ranges from 82% to 104%. This method was used for the quantification of pesticides in sugar samples collected from different factory outlets from different parts of India. In this study, 27 refined sugar samples were analysed in which one sample showed a detectable level of the chlorpyrifos. This study showed that Indian sugar is free from the commonly-used pesticides at the low μgg^{-1} levels (Sinha et al., 2011).

Pesticide	Matrices	Sample preparation	Instru-ment used	[a]LOD ranged (µgkg-1)	[b]LOQ ranged (µgkg-1)	% Recovery ranged (%mean)	[c]RSD ranged (%)	Refe-rences
OPs	vegetables	Solvent extraction	GC-PID	3 to10	NR	75.2% to 111.5%,	2.8% to 10.4%.	Li et al., 2010
OPs	vegetables	solvent extraction with SPE	CEI with quantum dot fluore-scence detection	50 to 180, and 100-30000	NR	88.7-96.1%		Chen et al., 2010)
pesticides	Fruit+ vegetable	QuEChERS	LC-MS/ MS	NR	NR	70% to 120%,	>20%	Camino-Sánchez et al., 2010)
pesticides	olives	solid-phase dispersion and QuEChERS	LC-MS/ MS	10	NR	70-120%		Gilbert-López et al., 2010
Pesticides (OCs + Ops)	vegetable	QuEChERS	GC-NPD	1 to 9	NR	70.22% to 96.32%	>15%	Srivastava et al., 2010
pesticides	fruits and vegetables	QuEChERS	LP-GC/ TOFMS	NR	NR	70-120%	< 20%	Koesukwiwat et al., 2010
Pesticides	vegetables	NR	GC-MS	NR	NR	NR	NR	Osman et al., 2010
Pesticides	vegetables	QuEChERS	LC-MS/MS	NR	NR	NR	NR	Kmellár et al., 2010
Ops and carba-mates	total diet	QuEChERS	LC-MS/ MS	NR	200	70-120%	< 20%	Chung et al., 2010
pesticides	vegetables	Liquid extraction	HPLC	0.5-3.0	NR	77.8-98.2%	NR	Lin et al., 2010
pesticides	Vegetable + fruits	Solid phase extraction	UPLC-MS/MS	1-2000	NR	70-120%	NR	Kamel et al., 2010

NR = Not reported
LOD= Limit of detection
LOQ= Limit of quantification
RSD= Relative standard deviation

Table 1. Different reported methods in food matrices.

The reported method for the analysis of pesticides in food samples is tabulated in Table-1. However, recently, a method has been reported (Li Wu, Chen & Zhang 2010) detailing the percentage recoveries, RSD, and LOD in vegetable sample. The reported RSD for all the pesticides used in this study ranged from 2.8% to 10.4% and the recovery ranged from 75% to 11%. The reported LOD in vegetable samples ranged from 3 to 10μgkg^{-1} for all pesticides. Also, an LC-MS/MS method has been used for the analysis of pesticides in fruit and vegetable samples (Camino sanchez et al., 2010). In this method, QuEChERS was used for the extraction. The reported mean percentage recoveries mostly ranged between 70% and 120%, and RSD were generally below 20% at the lower spiked level. The CEI with quantum dot fluorescence detector was used for analysis of vegetable samples (Chen & Fung 2010). The reported LOD ranged from 50 to 180μgkg^{-1}, while the reported recovery ranged between 88% and 96%. Additionally, the LC-MS/MS was used by different researcher (Gilbert-Lopez et al., 2010; Kmellar, Abranko, Fodor & Lehotay 2010; Chun & Chan 2010) for analysis of pesticides in olives, vegetables and total diet samples. In this method, QuEChERS was used for the extraction. The reported mean percentage recoveries mostly ranged between 70% and 120%, and RSD were below 20%. The reported LOD was 10μgkg^{-1}. The GC-MS method was used for analysis of pesticides in fruits and vegetables samples (Koesukwiwat, Lehotay, Miao & Lepipatpiboon, 2010; Osman, Al-Humaid, Al-Rehiayani & Al-Redhalman, 2010). The simultaneous identification and quantification of pesticide residues in fruits, milk and vegetables were reported (Kamel, Qian, Kolbe & Stafford, 2010). This method was based on solid phase extraction followed by UPLC-MS/MS chromatographic separation and a full-scan mass spectrometric detection. The recovery for all pesticides ranged from 70% to 120% with LOD ranged from 1 to 2000μgkg^{-1}. The HPLC method was also used for analysis of pesticides in vegetables samples (Lin et al., 2010). In this method liquid extraction was used. In the vegetable samples, the observed recoveries ranged from 77% to 99% and LOD ranged from 0.5 to 3μgkg^{-1} while GC-NPD was used for analysis of vegetable samples reported mean percentage recoveries mostly ranged between 70% and 96%, and RSD were generally below 15% (Srivastava, Trivedi, Srivastava, Lohani & Srivastava 2010). The LOD values ranged from 1 to 9μgkg^{-1}. However it should be noted that HPLC and GC-NPD are not confirmatory techniques.

4. Gas chromatography equipped with different detector for analysis of pesticides in biological samples

A sensitive GC-MS method in MS/MS ion preparation was developed for quantitative estimation and qualitative determination of chlorpyrifos in human blood samples. Dissociation energy effects on ion formation of chlorpyrifos and sensitivity of this analytical method was well demonstrated. Chlorpyrifos was extracted using methanol/hexane mixture from 0.2 ml human blood, deactivated with saturated acidic salt solution. The extract was then re-concentrated and analyzed by electron impact (EI) MS/MS gas chromatography-mass spectrometer. The MS/MS spectra of chlorpyrifos ion were recorded on different dissociation energy (30–100 V) to establish the structural confirmation and to demonstrate the effect of this energy on sensitivity, S/N ratio and detection limit for quantification of chlorpyrifos, which is reported for the first time. At different exciting amplitude (30–100 V), different behaviors of base peak, sensitivity, S/N ratio and detection limit of this method were observed for quantification of chlorpyrifos. The mass spectra recorded at dissociation energy <70 V, in between 70–80 V and >80 V correspond to the m/z 314 (100%), m/z 286 (100%) and m/z 258 (100%), respectively. The sensitivity, signal to noise ratio and detection limit of this method increased on 95 V at m/z 258. Therefore, m/z 258 was used for the quantification of chlorpyrifos. The detection limit for quantification was 0.1 ng/ml with S/N: 2 in human blood. The linear calibration curve with the correlation coefficient ($r > 0.99$) was obtained. The percentage recoveries from 95.33% to 107.67% were observed for chlorpyrifos from human blood. The blood samples were collected at different time intervals. The concentration of chlorpyrifos in Poisoning case was 3300, 3000, 2200, 1000, 600, 300 ng/ml on day 1, 3, 6, 8, 10 and 12, respectively. On the 12th day of exposure of chlorpyrifos, 90.9% reduction in concentration was observed. On day 14th, the chlorpyrifos was not detected in the blood of the same subject. Thus, the present study is useful for detection of chlorpyrifos in critical care practices and also provides tremendous selectivity advantages due to matrix elimination in the parent ion isolation step in blood sample analysis for chlorpyrifos (Sinha et al., 2006). The effect of dissociation energy on ion formation and sensitivity of triazofos in blood samples was studied. Six millilitres hexane was used for the extraction of triazofos from 2mL serum samples. The extract was reconstituted in 1mL hexane and analyzed by GC-MS/MS in electron impact MS/MS mode. The structure, ion formations, nature of base peak and fragmentation schemes were correlated with the different dissociation energies. The new ion was obtained at mass to charge ratio 161 (100%), which was the characteristic ion peak of triazofos. On using different exciting amplitudes, different behaviours of fragmentation schemes were obtained. The effect of dissociation energy on sensitivity of the analyte was also demonstrated. The mass spectra recorded at different exciting amplitudes <50V in between 50-60V and >60V correspond to the m/z 161 (100%), 77 (100%) and 119 (100%), respectively. The maximum sensitivity of analyte in blood sample was obtained on using 50V dissociation energy. Additionally, the effect of current on sensitivity of the method was also demonstrated. In all conditions, the new characteristic ion at m/z 161 was obtained and used for quantification of triazofos in blood samples with maximum sensitivity. The limit of detection and quantification was 0.351 and 1.17 ngmL^{-1}, respectively, with 99% accuracy. The observed correlation coefficient was 0.995. The inter-day percentage recoveries from 83.9% to 111% were obtained below 9.38 percentage RSD. This present method gives combined picture of confirmation and quantification of triazofos in critical care practices and also provides tremendous selectivity advantages due to matrix elimination in the parent ion isolation step in blood sample analysis for triazofos in poisoning emergency cases (Sinha 2010). A number of families in a rural area of Jabalpur District (Madhya Pradesh), India, were affected by repeated episodes of convulsive illness over a period of three weeks. The aim of this investigation was to determine the cause of the illness. The investigation included a house-to-house survey, interviews of affected families, discussions with treating physicians, and examination of hospital records. Endosulfan poisoning was suspected as many villagers were using empty pesticide containers for food storage. To confirm this, our team collected blood and food samples, which were transported to the laboratory and analyzed with GC-ECD. Thirty-six persons of all age groups had illness of varying severity over a period of three weeks. In the first week, due to superstitions and lack of treatment, three children died. In the second week, symptomatic treatment of affected persons in a district hospital led to recovery but recurrence of convulsive episodes occurred after the return home. In the third week, 10 people were again hospitalized in a teaching hospital. Investigations carried out in this hospital ruled out infective etiology but no facilities were available for chemical analysis. All persons responded to symptomatic treatment. The presence of endosulfan in blood and food samples was confirmed by GCMS. One of the food items (Laddu) prepared from wheat flour was found to contain 676 ppm of α-endosulfan. Contamination of wheat grains or flour with endosulfan and its consumption over a period of time was the most likely cause of repeated episodes of convulsions, but the exact reason for this contamination could not be determined. This report highlights the unsafe disposal of pesticide containers by illiterate farm workers, superstitions leading to delay in treatment, and susceptibility of children to endosulfan (Dewan et al., 2004). Mass spectrometry method has been used with capillary gas chromatography in electron ionization mode to identify the presence of α, β and endosulfan sulphate in human serum samples. The fragmentation ions were obtained due to ring cleavages, rearrangement and remote charge process. Solvent extraction procedure was used for isolation of compounds from serum sample. A detection limit as low as 100 ppb for α, β and endosulfan sulphate could be easily achieved for confirmation. For endosulfan and endosulfan sulphate the m/z values were obtained at intervals of M^{+2}. The m/z values of α, β endosulfan are identical but they differed in retention time (Sinha et al., 2004). Therefore, the spectra reported might serve as reference spectra for identification of different type of organo-chlorine pesticide in human serum samples. The GC-MS has been employed to quantitatively detect at low ppb levels of α and β endosulfan in human serum, urine and liver but this failed to separate α, β-isomers. GC equipped with a mass detector was used to measure the levels of endosulfan in human blood at ppb level.

Although some OC pesticide metabolites are monitored in urine, they are most commonly measured as the intact pesticide and/or its metabolite in whole blood, serum, plasma, or other lipid-rich matrices. These methods and the specific pesticides measured are outlined in Table 2. Although some pesticide metabolites are monitored in urine, they are most commonly measured as the intact pesticide and/or its metabolite in, whole blood, serum, plasma, or other lipid-rich matrices. Typically, serum or plasma is extracted using a liquid partitioning or solid-phase extraction (SPE) and the extract is analyzed using capillary gas chromatography (GC) with electron-capture detection (ECD) (Table-2). These methods are reliable and use affordable instrumentation. However, GC–ECD analyses have a higher potential for detecting interfering components than do more selective analysis techniques. Other methods for analysis of serum extracts include mass selective detection (MSD aka MS) [Table-2] and high-resolution mass spectrometry with isotope dilution quantification. These analyses are typically more selective and sensitive than GC–ECD analyses; however, the high cost of instrumentation and isotopically labeled standards and the complex operation and maintenance of these instruments often preclude their routine use in most laboratories.

Typically, serum or plasma is extracted using a liquid partitioning or solid-phase extraction (SPE) and the extract is analyzed using capillary gas chromatography (GC) with electron-capture detection (ECD) (Table-2). A sensitive and accurate analytical method was developed for quantifying 29 contemporary pesticides in human serum or plasma. These pesticides include organophosphates, carbamates, chloroacetanilides, and synthetic pyrethroids among others and include pesticides used in agricultural and residential settings. This method employs a simple solid-phase extraction followed by a highly selective analysis using isotope dilution gas chromatography–high-resolution mass spectrometry. This method is very accurate, has limits of detection in the low pg/g range and coefficients of variation of typically less than 20% at the low pg/g end of the method linear range (Barr et al., 2002).

Method	Analytes[a]	Matrix	Extraction	Analytical system	I.S. type	Recovery[e] (%)	LOD (μg/g)	RSD (%)
Frenzel, 2000	15	Whole blood	Kieselguhr SPE	GC–MS	None	97	30–40	7
Rohrig, 2000	1–6	Breast milk	SPME	GC–ECD	None	b	Low	b
Ward, 2000	1–10	Serum, breast	SPE	GC–HRMS	Isotope dilution	60–80	0.07–0.26	<20
Najam, 1999	1–3, 5–12, 13, 17–18	Serum	Solvent extraction, Silica / Florosil cleanup	GC–ECD	Surrogates	39–126	0.15–0.5	7–32
Pauwels, 1999	f	Serum	C_{18} SPE, acid wash	GC–MS	$^{13}C_{12}$ PCB 149	48–140	Low	b
Lino, 1998	f	Serum	Florisil SPE	GC–ECD	f	>84	1–37 μg/l	<!9
Luo, 1997	1,2	Serum	*n*-Hexane	GC–ECD	b	93–106	b	b
Walisze-wski, 1982	1, 2,4,6	Adipose	Light petroleum, acid wash	GC–ECD	b	91–99	0.01 μg/kg	<10
Brock, 1996	1–11	Serum	C_{18} SPE, Florosil	GC–ECD	Surrogates	63–80	0.08–0.66 μg/l	0.7–5.9[c]
Noren, 1996	MeSO - DDE	Breast milk	Liquid–gel partitioning, absorption/ gel permeation chromatography	GC–HRMS	Surrogates	80–97	0.01–0.05 ng/g lipid	4–14
Gill, 1996	1–9, 11–13, 15+ others	Serum	Solvent extraction /SPE	GC–MS	Surrogates	60–110	b	b
Guardino, 1996	1, 2	Blood	C_{18} SPE	GC–ECD;	None	b	b	b
Minelli, 1996	1, 2, 4–6	Serum	Serum–silica suspension, hexane/ acetone, alumina cleanup	GC–ECD	b	80–99	< 1 mg/l	< 15
Gallelli, 1995	f	Adipose liver	Light petroleum, Florosil	GC–ECD	f	f	f	f

Method	Analytes[a]	Matrix	Extraction	Analytical system	I.S. type	Recovery[e] (%)	LOD (µg/g)	RSD (%)
Prapamontol, 1991	1–2, 4–7, 10–13	Milk	Ethyl acetate / acetone /methanol, C_{18}SPE	GC–ECD	Surrogates	90–110	0.5–2.5 µg/l	≤16
Buroc, 1990	1–3, 5–10, 12	Serum	Hexane/ ether, Florosil	GC–ECD	Surrogates	48–122	<1 µg/l	7–23
Saady, 1990	1–2, 4, 6, 7–10	Serum	C_{18} SPE	GC–ECD	Surrogate	70–75	0.1–0.7 µg/l	4–25
Liao, 1988	1–5, 11	Adipose	solvent, Florosil	GC–MS	Surrogat	72–120	b	b
Mardones 1999	14	Urine	Acid hydrolysis, on-line cleanup SFE	MEKC–UV	b	58–103	1–12 µg/l	3–7
LeBel, 1983	f	f	Acetone/ hexane, gel permeation,. dichloromethane/ cyclohexane	GC–ECD	B	>80	b	b
Bristol, 1982	3, 5, 7–9, 11	f	f	GC–ECD[d]	Surrogates	35–99	Low µg/ l	3–20
Tessari, 1980	1–3, 7–10, 13	Breast milk	ACN/ hexane, Florosil	GC–ECD	b	68–90	0.5–30 µg/l	b
Strassman 1977	1–13	Breast milk	Solvent, Florosil	GC–ECD[d]	b	b	10–100 µg/l	b
Martinez, 1998	15	Urine	SPE	GC–MS–MS	Dieldrin	>89	0.006–0.018 µg/l	9–13
Nigg, 1991	17	Urine	Oxidation, solvent	GC (detector not noted)	None	b	1	b
Angerer, 1981	14	Urine	Acid hydrolysis, derivatization	GC–ECD	f	87–119	5–20 µg/l	4–10
Holler, 1989	14	Urine	Acid hydrolysis, derivatization	GC–MS–MS	Isotope analogues/surrogates	> 50	1	c
Hill, 1995	14	Urine	Enzyme hydrolysis, chlorobutane / ether; derivatization	GC–MS–MS	Stable isotope analogues	b	1–2 µg/l	21–24

Method	Analytes[a]	Matrix	Extraction	Analytical system	I.S. type	Recovery[e] (%)	LOD (μg/g)	RSD (%)
Heleni Tsoukali et al., 2005	malathion, parathion, methyl parathion and diazinon	bio-logical samples	SPME	GC-NPD	Fenitro-thion	b	2 to 55 ng/ml	b
Zlatković M, et al., 2010	dimethoate, diazinon, malathion and malaoxon	bio-logical samples	solid-phase extraction	LC-MS	b	90-99	0.007-0.07 mg/l	b
John H et al., 2010	dimethoate	bio-logical samples	Deproteini-zation by precipitation and extensive dilution	LC-MS/MS	b	90-115	0.12-0.24 μg/l	5-14
Paula Proença et al., 2005	imi-dacloprid	bio-logical samples	b	LC-MS	b	86	0.002 μ g/mL	5.9
Gallardo et al., 2006	dimethoate	bio-logical samples	SPME	GC-MS	ethion	1.24 and 0.50%.	50-500 ng/ml	
Saito et al., 2008		bio-logical samples	SPME	HS-GC-MS	b	b	0.25 μg/mL	12.6
Bichon et al 2008	Fipronil	bio-logical samples	SPE	GC-MS	b	b	0.05-0.73 pg/ml	b
Zhou et al., 2007	pentachlor ophenol	bio-logical samples	HS-SPME	GC-MS	b	b	0.02 ng/ml	12.6

I.S.= Internal standard; LOD = limit of detection; RSD = relative standard deviation; SPME = solid-phase micro-extraction; SPE = solid-phase extraction; SFE = supercritical fluid extraction; GC–ECD = gas chromatography electron capture detection; GC–MS= gas chromatography–mass spectrometry, GC–HRMS=gas chromatography–high-resolution mass spectrometry;. a 1= *p*, *p*-DDT; 2= *p*, *p*-DDE; 3= hexachlorobenzene; 4= a-hexachlorocyclohexane; 5= b-hexachlorocyclohexane; 6= g-hexachlorocyclohexane; 7= heptachlor epoxide; 8= oxychlordane; 9= *trans*-nonachlor; 10= dieldrin; 11= aldrin; 12= endrin; 13= mirex; 14= lindane and/or metabolites (chlorinated phenols); 15= endosulfan and/or metabolites; 16= methylsulfonyl-DDE; 17= *o*, *p*-DDT; 18=*cis*-nonachlor. b Not given. c Standard error about the mean. d GC–MS used for confirmation of positive samples. e Recovery refers only to the absolute recovery from extraction or isolation of the analyte. f Unable to obtain full article. Details taken from abstract. Missing details may be available in full article.

Table 2. Methods for analysis of organochlorine pesticides in human matrices

5. Conclusion

In this chapter, we present a review of existing analytical methodology for the biological and environmental monitoring of exposure to pesticides has been presented. A critical assessment of the existing methodology has been done and other areas which need more research have been identified. An effort has been made to use to measure toxicants. There are many detectors which are equipped with Gas Chromatography like Mass, electron capture detector, flame ionization detector, phosphorous nitrogen detector etc. But, in one modern context the mass detector is the best detector which is very useful for confirmation and quantification of all types of compounds. The electron capture detector is only suitable for quantification of organo-chlorine compounds but it is not confirmatory technique. The nitrogen phosphorous detector is suitable for quantification of nitrogen and phosphorous compounds. Other types of detectors except mass detector are not suitable as confirmatory technique.

6. References

A.R., Najam, M.P., Korver, C.C., Williams, V.W., Burse, & L.L., Needham, (1999). Analysis of mixture of polychlorinated biphenyl and chlorinated pesticides in human serum by column fraction and dual-column capillary gas chromatography with electron capture detection. J. AOAC Int. 82, 177-185.

Abdulrahman, R. B. (2001). Distribution of VOC in ambient air of Tehran , Arch Env Health, 56, 380-383,

Angerer, J., Heinrich, R., & Laudehr, H., (1981). Occupational exposure to hexachlorocyclohexane. V. Gas chromatographic determination of monohydroxychlorobenzenes (chlorophenols) in urine. Int Arch Occup Environ Health. 48(4), 319-24

Arrebola, F.J., Martinez Vidal, J.L., & Fernandez-Gutierrez , A., (1999). Excretion study of endosulfan in urine of a pest control operator. Toxicology Letters, 107, 15-20.

Ballschmiter, R., Bacher, A., Mennel, R.C., Fischer, U., Riehle, Swerev, M., (1992). The determination of chlorinated biphenyls, chlorinated dibenzodioxins and chlorinated dibenzofurans by GC-MS. J high Resolution chromatography 15, 260-270. Barcelona. Hum Toxicol 6, 397-400

Barr, D. B., Barr, J. R., Maggio, V. L., Whitehead, R. D., Jr., Sadowski, M. A., Whyatt, R., et al. (2002). A multi-analyte method for the quantification of contemporary pesticides in human serum and plasma using high-resolution mass spectrometry. Journal of Chromatography B, 778, 99–111.

Bichon, E., Richard, C.A., & Le Bizec, B., (2008). Development and validation of a method for fipronil residue determination in ovine plasma using 96-well plate solid-phase extraction and gas chromatography-tandem mass spectrometry, J Chromatogr A. 2008 1201(1), 91-9.

Bristol, D.W., Crist, H.L., Lewis, R.G., MacLeod, K.E., Sovocool, G.W., (1982). Chemical analysis of human blood for assessment of environmental exposure to semivolatile organochlorine chemical contaminants, J Anal Toxicol. 6(6), 269-75.

Brock, J.W., Burse, V.W., Ashley, D.L., Najam, A.R., Green, V.E., Korver, M.P., Powell, M.K., Hodge, C.C., & Needham, L. L., (1996). An improved analysis for chlorinated pesticides and polychlorinated biphenyls (PCBs) in human and bovine sera using solid-phase extraction. J Anal Toxicol 20, 528–536.

Burse, V.W., Head, S.L., Korver, M.P., McClure, P.C., Donahue, J.F., & Needham, L.L., (1990). Determination of selected organochlorine pesticides and polychlorinated biphenyls in human serum, J Anal Toxicol. 1990 May-Jun;14(3):137-42

C. Mardones, A., Rios, M., & Valcarcel, (1999). Determination of chlorophenols in human urine based on the integration of on-line automated clean-up and preconcentration unit with micellar electrokinetic chromatography, Electrophoresis, 20, 2922-2929.

Camino-Sánchez, F.J., Zafra-Gómez, A., Oliver-Rodríguez, B., Ballesteros, O., Navalón, A., Crovetto, G., Vílchez, J.L., (2010) UNE-EN ISO/IEC 17025:2005-accredited method for the determination of pesticide residues in fruit and vegetable samples by LC-MS/MS. Food Addit Contam Part A Chem Anal Control Expo Risk Assess. 27(11), 1532-44

Carmichael, P.L., Jacob, J., Grimmer, G., Phillips, D.H. (1990). Analysis of the PAH content of petrol and diesel engine lubricating oils and determination of DNA adductsin topically treated mice by 32 P-post-labelling, Carcinogenesis, 11, 2025-2032

Chen, Q. Fung, Y., (2010). Capillary electrophoresis with immobilized quantum dot fluorescence detection for rapid determination of organophosphorus pesticides in vegetables. Electrophoresis, 31 (18), 3107-3114.

Chung, S.W., Chan, B.T., (2010). Validation and use of a fast sample preparation method and liquid chromatography-tandem mass spectrometry in analysis of ultra-trace levels of 98 organophosphorus pesticide and carbamate residues in a total diet study involving diversified food types, J Chromatography A, 1217(29), 4815-24.

Dewan, A., Bhatnagar, V. K., Mathur, M. L., Chakma, T., Kashyap, R., Sadhu, G. H., et al. (2004). Repeated episodes of endosulfan poisoning. Journal of Toxicology –Clinical Toxicology, 42, 1–7.

E. V., Minelli & M. L., Ribeiro, (1996). Quantitative method for the determination of organochlorine pesticides in serum. *J. Anal. Toxicol.*, 20(1), 23-26.

Fischer, R. C., Witlinger, R., Ballschmiter, K., (1992). Retention index based vapour pressure estimation for PCB by GC, Fresenius. J of Analytical Chemistry 342, 421-425

Frenzel, T. Sochor, H., Speer K., & and Uihlein, M., (2000). Rapid multi-method for verification and determination of toxic pesticides in whole blood by means of capillary GC-MS. *J. Anal. Toxicol.*, 24(5), 365-371.

Frenzel, T., Sochor, H., Speer, K., & Uihlein, M., (2000) Rapid multi-method for verification and determination of toxic pesticides in whole blood by means of capillary GC-MS. *J. Anal. Toxicol.*, 24(5), 365-371

Gallardo, E., Barroso, M., Margalho, C., Cruz, A., Vieira, D. N., & López-Rivadulla, M., (2006). Solid-phase microextraction for gas chromatographic/mass spectrometric analysis of dimethoate in human biological samples, Rapid Communications in Mass Spectrometry 20(5), 865–869.

Gallelli, G., Mangini, S.,& Gerbino, C.,(1995). Organochlorine residues in humans adipose and hepatic tissues from autopsy sources in northern Italy. J. Toxicol. Environ Health 46, 293-300.

Gomez-Catalan, J. J., To-Figueras, J., Planas, M., Rodamilans, J., Corbella, (1987).

Guardino, X., Serra, C., Obiols, J., Rosell, M.G., Beren guer, M.J., Lopez, F., & Brosa, J., (1996). Determination of DDT and related compounds in blood samples from agricultural workers J. Chromatogr. A 719 , 141-147.

Heleni, T., Georgios, T., Nikolaos, R., & Ifigeneia G., (2005). Solid phase microextraction gas chromatographic analysis of organophosphorus pesticides in biological samples , Journal of Chromatography B, 823 (1 2)194-200.

Hill, R.H., Shealy Jr, D.B., Head, S.L., Williams, C.C., Bailey, S.L., Gregg, M., Baker, S.E., & Needham, L.L., (1995). Determination of Pesticide Metabolites in Human Urine Using an Isotope Dilution Technique and Tandem Mass Spectrometry J. Anal. Toxicol. 19, 323-329.

Hodgson AT. Grimmer JR. Application of multisorbent sampling technique for investigations of volatile organic compounds buildings, In: Design and protocol for monitoring of indoor air-quality. ASTM STP 1002, NL Nagota and JP Harper (Eds.) 1989, 244-256

Holler J.S., Fast, D.M., Hill, R.H., Jr, Cardinali, F.L., Todd, G.D., McCraw, J.M., Bailey, S.L., & Needham, L.L., (1989). Quantification of selected herbicides and chlorinated phenols in urine by using gas chromatography/mass spectrometry/mass spectrometry. J Anal Toxicol. 13(3),152-7.

John, H., Eddleston, M., Clutton, R.E., Worek, F., & Thiermann, H., (2010). Simultaneous quantification of the organophosphorus pesticides dimethoate and omethoate in porcine plasma and urine by LC-ESI-MS/MS and flow-injection-ESI-MS/MS, J Chromatogr B Analyt Technol Biomed Life Sci. 878(17-18), 1234-45

Kamel, A., Qian, Y., Kolbe, E., Stafford, C., (2010). Development and validation of a multiresidue method for the determination of neonicotinoid and macrocyclic lactone pesticide residues in milk, fruits, and vegetables by ultra-performance liquid chromatography/MS/MS. J AOAC International, 93(2), 389-99.

Keenan, F., Doug, C., Vegetation clearance study work plan ordinance and explosive. Prepared by US Dept. of the Army Sacrament District Corps of Engineers, 1325, J Street, Sacramento, California 95814-2922. HLLA project no. 46310001131

Kmellár, B., Abrankó, L., Fodor, P., Lehotay, S.J., (2010), Routine approach to qualitatively screening 300 pesticides and quantification of those frequently detected in fruit and vegetables using liquid chromatography tandem mass spectrometry (LC-MS/MS). Food Addit Contam Part A Chem Anal Control Expo Risk Assess. 27(10),1415-30.

Koesukwiwat , U., Lehotay, S.J., Miao, S., Leepipatpiboon, N., (2010) High throughput analysis of 150 pesticides in fruits and vegetables using QuEChERS and low-pressure gas chromatography-time-of-flight mass spectrometry. J Chromatography A, 1217(43), 6692-703.

Kumari, B., & Kathpal, T. S. (2009). Monitoring of pesticide residues in vegetarian diet. Environmental Monitoring Assessment, 151, 19–26.

Kumari, B., Madan, V. K., & Kathpal, T. S. (2006). Monitoring of pesticide residues in fruits. Environmental Monitoring and Assessment, 123, 407–412.

Kumari, B., Madan, V. K., Kumar, R., & Kathpal, T. S. (2002). Monitoring of seasonal vegetables for pesticide residues. Environmental Monitoring and Assessment, 74, 263–270.

LeBel G.L., & Williams, D.T., (1983). Determination of organic phosphate triesters in human adipose tissue. J Assoc Off Anal Chem. 66(3), 691-9

Li, L., Wu, C., Chen, J., Zhang, S., Ye, Y., (2010). Determination of organophosphorus pesticide residues in vegetables by gas chromatography using back-flush technique. Se Pu. July 28 (7), 724-8

Liao, W., Smith, W.D., Chiang, T.C., & Williams, L.R., (1988). Rapid, low-cost cleanup procedure for determination of semi-volatile organic compounds in human and bovine adipose tissues , J Assoc Off Anal Chem.71(4), 742-7

Lin, X., Chen, X., Huo, X., Yu, Z., Bi, K., Li, Q., (2010). Dispersive liquid-liquid microextraction coupled with high-performance liquid chromatography-diode array detection for the determination of N-methyl carbamate pesticides in vegetables. journal of Separation Science. PMID: 21137098

Lino, C.M., Azzolini, C.B., Nunes, D.S., Silva, J.M., & da Silveira, M.I., (1998). Methods for the determination of organochlorine pesticide residues in human serum J. Chromatogr. B 716 , 147-152.

Luo, X.W., Foo, S.C., & Ong, H.Y., (1997). Serum DDT and DDE levels in Singapore general population Sci. Total Environ. 208 , 97-104.

Martínez Vidal, J.L., Arrebola, F.J., Fernández-Gutiérrez, A., & Rams, M.A., (1998). Determination of endosulfan and its metabolites in human urine using gas chromatography-tandem mass spectrometry, J Chromatogr B Biomed Sci Appl. 719(1-2), 71-8.

Martinez, J. L. V., Arrebola, F. J., Garrido, A. F., Martinez, J. F., & Mateu-Sanchez, M. (2004). Validation of a gas chromatographic–tandem mass spectrometric method for analysis of pesticide residues in six food commodities. Selection of a reference matrix for calibration. Chromatographia, 59(5–6), 321–327.

Miller, K. D., & Milne, P. (2008). Determination of low-level pesticides residues (0.5 lg/L) in soft drinks and sport drinks by gas chromatography with mass spectrometry: Collaborative study. Journal of AOAC International, 91(1), 202–236.

Najam, A.R., Korver, M.P., Williams, C.C., Burse, V.W., & Needham, L L., (1999) Analysis of a mixture of polychlorinated biphenyls and chlorinated pesticides in human serum by column fractionation and dual-column capillary gas chromatography with electron capture detection. J. AOAC Int. 82, 177-185.

Nigg, H. N., Stamper, J. H., Deshmukh S. N., & Queen, R. M., (1991). 4,4'-Dichlorobenzilic acid urinary excretion by dicofol pesticide applicators, Chemosphere. 22, (3-4) , 365-373.

Norén, K., Lundén, A., Pettersson, E., & Bergman, A., (1996). Methylsulfonyl metabolites of PCBs and DDE in human milk in Sweden, 1972-1992. Environ Health Perspect. 104(7), 766–772

Osman, K.A., Al-Humaid, A. M., Al-Rehiayani, S.M., Al-Redhaiman, K.N., (2010). Monitoring of pesticide residues in vegetables marketed in Al-Qassim region, Saudi Arabia. Ecotoxicol Environ Saf. 73(6), 1433-9.

Paula, P., Helena, T., Fernando, C., João, P., Paula ,V., Monsanto, E., Marques, P., & Duarte N. V., (2005). Two fatal intoxication cases with imidacloprid: LC/MS analysis, Forensic Science International. Volume 153(1) ,75-80

Pauwels, A., Wells, D., Covaci, A., & Schepens, P.J.C. (1999). Improved sample preparation method for selected persistent organochlorine pollutants in human serum using solidphase disk extraction with gas chromatographic analysis. J Chromatogr B 723:117-125.

Paya, P., Anastassiades, M., Mack, D., Sigalova, I., Tasdelen, B., Oliva, J., et al. (2007). Analysis of pesticide residues using the Quick Easy Cheap Effective Rugged and Safe (QuEChERS) pesticide multiresidue method in combination with gas and liquid chromatography and tandem mass spectrometric detection. Analytical Bioanalytical Chemistry, 389, 1697–1714.

Pentachlorophenol and hexachlorobenzene in serum and urine of the population of Prapamontol T., & Stevenson, D., (1991). Rapid method for the determination of organochlorine pesticides in milk. J Chromatogr 552:249-257.

Quach, D.T.I.,. Ciszkouzski, N. A. P., Fienlayson, J., Barbara, A., (1998). New GC-MS experiment for the undergraduate instrumental laboratory in environmental chemistry: methyl –t-butyl ether and benzene in gasoline. J Chem Educ 75. 595-599.

Reidel, K. (1996). Determination of benzene and alkylated biomass in ambient and exhaled air by microwave desorption coupled with GC-MS. J of Chromatography A 719, 383-389.

Robert H. TPH measurements; the advantage of using GC-MS. 1995: PhD. Dissrtation. ENTRIX INC JHON MAC MURPHY ZYMAX ENVIR TECHNOLOGY INC.

Rohrig, L., Meisch, H.U., & Fresenius, J., (2000). Application of solid phase micro extraction for the rapid analysis of chlorinated organics in breast milk Anal. Chem. 366 , 106-111

Saady J.J., & Poklis A., (1990). Determination of chlorinated hydrocarbon pesticides by solid-phase extraction and capillary GC with electron capture detection J Anal Toxicol. 1990 Sep-Oct;14(5):301-4

Saito, T., Morita, S., Motojyuku, M., Akieda, K., Otsuka, H., Yamamoto, I., & Inokuchi, S., (2008). Determination of metaldehyde in human serum by headspace solid-phase microextraction and gas chromatography-mass spectrometry, J Chromatogr B Analyt Technol Biomed Life Sci. 875(2), 573-6.

Sinha SN. Kulkarni, P.K. Desai, N. M., Shah, S. H., Patel, G.M., Mansuri, M.M., Parih, D. J., Saiyed, H.N. (2005). Gas chromatographic-mass spectroscopic determination of benzene in Indoor air during the use of biomass fuels in cooking time, J of chromatography A 315-319.

Sinha SN. Kulkarni, P.K. Shah, S. H., Desai, N. M., Patel, G.M., Mansuri, M.M., Saiyed, H.N. (2006). Environmental monitoring of benzene and toluene produced in indoor air due to combustion of solid biomass fuels, Science of the Total Environment 357, 280-287.

Sinha, S. N. (2010). Effect of dissociation energy: Signal to noise ratio on ion formation and sensitivity of analytical method for quantification and confirmation of triazofos in blood samples using gas chromatography–mass spectrometer (GC–MS/MS). International Journal of Mass Spectrometry, 296, 47–52.

Sinha, S. N., Bhatnagar, V.K., Doctor, P., Toteja, G.S., Agnihotri, N.P., Kalra R.L., A novel method for pesticide analysis in refined sugar samples using a gas chromatography-mass spectrometer (GC-MS/MS) and simple solvent extraction method, Food Chemistry 126 (2011) 379–386

Sinha, S. N., Pal, R., Dewan, A., Mansuri, M. M., & Saiyed, H. N. (2006). Effect of dissociation energy on ion formation and sensitivity of an analytical method for determination of chlorpyrifos in human blood, using gas chromatography-masss pectrometer (GC-MS in MS/MS). International Journal of Mass Spectrometry, 253, 48–57.

Sinha, S. N., Patel, T. S., Desai, N. M., Mansuri, M. M., Dewan, A., & Saiyed, H. N. (2004). GC-MS study of endosulfan in biological samples. Asian Journal of Chemistry, 16(3-4), 1685-1690.

Sinha, S. N., Zadi, S. S. (2008). Gas Chromatography-Mass spectral analysis of paint thinner, Asian Journal of Chemistry 20(8), 6365—6368.

Srivastava, A.K., Trivedi, P., Srivastava, M.K., Lohani, M., & Srivastava, L.P. (2011). Monitoring of pesticide residues in market basket samples of vegetable from Lucknow City, India: QuEChERS method. Environmental Monitoring and Assessment, 176(1-4, 465-472.

Star J. M., Rone, J.R., Metnacche, M.G., Henderson, R. F. (2000). Development of methods for analysis of biomarkers of Benzene exposure using GC/MS and MALDI. The Toxicologist. Presented at the Annual meeting of the society of Toxicology, Philadelphia, 54, 310.

Strassman S.C., & Kutz, F.W., (1977). InseAicide residues in human milk from Arkansas and Mississippi,1 973-1974.Pestic Monit J , 10(4),1 30-133.

Tessari, J.D., & Savage, E.P., (1980). Gas-liquid chromatographic determination of organochlorine pesticides and polychlorinated biphenyls in human milk. J Assoc Off Anal Chem. 1980 Jul;63(4):736-41.

U. S., Gill, H. M., Schwartz, B. Wheatley, (1996). Development of a method for the analysis of PCB congeners and organochlorine pesticides in blood/serum. Chemosphere Volume: 32 (6), 1055-1076.

Waliszewski, S.M., & Szymczynski, G.A., (1982). Simple, low-cost method for determination of selected chlorinated pesticides in fat samples J. Assoc. Off. Anal. Chem. 65, 677-9.

Ward, E.M., Schulte, P., Grajewski, B., Andersen, A., Patterson Jr., D.G., Turner, W., Jellum, E., Deddens, J.A., Fried- land, J., Roeleveld, N., Waters, M., Butler, M.A., DiPietro, E., & Needham, L.L., (2000). Serum organochlorine levels and breast cancer: a nested case-control study of Norwegian women. Cancer Epidemiol. Biomarkers Prev. 9, 1357-1367.

Ward, E.M., Schulte, P., Grajewski, B., Andersen, A., Patterson, D.G., Jr, Turner, W., Jellum, E., Deddens, J.A., Friedland, J., Roeleveld, N., Waters, M., Butler, M.A., DiPietro, E., & Needham, L.L., (2000). Serum organochlorine levels and breast cancer: a nested case-control study of Norwegian women. Cancer Epidemiol Biomarkers Prev. 9(12), 1357-67.

Zhou, Y., Jiang, Q., Peng, Q., Xuan, D., & Qu, W., (2007). Development of a solid phase microextraction-gas chromatography-mass spectrometry method for the determination of pentachlorophenol in human plasma using experimental design. Chemosphere. 2007 70(2), 256-62.

Zlatković, M., Jovanović, M., Djordjević, D., & Vucinić, S., (2010). Rapid simultaneous determination of organophosphorus pesticides in human serum and urine by liquid chromatography-mass spectrometry, Vojnosanit Pregl. 67(9),717-22.

Acetone Response with Exercise Intensity

Tetsuo Ohkuwa[1,*], Toshiaki Funada[1] and Takao Tsuda[2]
[1]*Department of Materials Science and Engineering, Nagoya Institute of Technology, Showa-ku, Nagoya,*
[2]*Pico-Device Co., Nagoya Ikourenkei Incubator, Chikusa-ku, Nagoya, Japan*

1. Introduction

Blood samples are commonly used to explain the mechanism of metabolism during exercise and diagnosis of certain diseases. However, blood sampling is invasive and usually accompanied by problems such as loss of blood, emotional stress and discomfort for volunteers. The collection and analysis of expired air and skin gas has a number of advantages compared to that of blood, and can be performed by non-invasive and painless procedure. These non-invasive volatile gas monitoring is attractive since they can be repeated frequently without any risk and their sampling does not require skilled medical staff. Especially, skin gases which emanated from human skin which enables collection from hands, arms, fingers and other local areas of the human bodies (Tsuda et al. 2011). Volatile organic compounds can be produced anywhere in the body and may reflect physiologic or pathologic biochemical processes. These substrates are transported via the bloodstream and exhaled through the lung (Schubert et al. 2004). Analysis of volatile organic compounds enable the observation of biochemical process in the body in noninvasive manner.

It well known that free fatty acid and glucose are the major energy fuels under usual circumstances. Ketones such as β-hydroxybutyrate, acetoacetate and acetone are generated in the liver via decarboxylation of excess Acetyl-CoA (Miekisch et al. (2004), mainly from the oxidation of fatty acids, and are exported to peripheral tissues, such as the brain, heart, kidney and skeletal muscle for use as energy fuels (Mitchell et al. 1995; Laffel et al. 1999). Acetone is one of the most abundant compounds in human breath (Miekisch et al. 2004). Acetone is mainly generated from non-enzymatic, decarboxylation of acetoacetate (Laffel et al. 1999; Owen et al. 1982). Ketones in blood increased gradually with respect to the length of fasting periods in diabetics and obese people (Tassopoulos et al. 1969; Reichard et al. 1979). Owen et al. (1982) demonstrated that in diabetic patients, plasma acetone concentration is significantly related to breath acetone concentration. Acetone in expired air increased after high-fat ketogenic diet treatment (Musa-Veloso et al. 2002). The relationship between acetone concentration in plasma and breath has been well established (Naioth et al. 2002). Yamane et al. (2006) reported that skin gas acetone concentrations of patients with diabetes were significantly higher than those of the control subjects. Turner et al. (2008) demonstrated that there is a clear relationship between the level of breath acetone and

* Corresponding Author

acetone released from the skin. This result suggested that emission rates of acetone from skin can be used to estimate the blood acetone level. The acetone concentrations in both skin and breath gases are potentially useful as an indicator of β-oxidation of fatty acid and ketosis (Naitoh et al. 2002; Miekish et al. 2004; Musa-Veloso et al. 2002), but less is known as to the effects of exercise on acetone concentration of skin gas and exhaled acetone.

Senthilmohan et al. (2000) reported that running exercise increased breath acetone concentration compared to the basal level. However, in their study exercise intensity was not clarified. Sasaki et al. (2011) reported that acetone concentration in expired air increased during graded and prolonged exercise such as walking or running. We have reported that acetone concentrations emanated from skin gas and in expired air increased with high exercise intensities of cycle exercise (Yamai et al. 2009) and hand-grip exercise (Mori et al. 2008).

2. Methods

2.1 Procedure of exercise

The participants for hand-grip and cycle exercise were 7 and 8 healthy males, respectively. When this study was started, the first step was to ascertain the reproducibility of the results for skin gas acetone concentration test during hand-grip exercise. First the skin gas acetone was collected for four times in one participant. In hand-grip exercise, the participants performed a dynamic hand-grip exercise of three different types of exercise during 60 sec (Exercise 1, 2 and 3). Exercise 1 was performed at 20 kg with one contraction per two sec. Exercise 2 was 30kg with one contraction per three sec. Exercise 3 was 10kg with one contraction per sec. In cycle exercise, the workloads were 360 (1.0kg), 720 (2.0kg), and 990 (2.75kg) kgm/min, and each stage was 5 min in duration. A pedaling frequency of 60 rpm was maintained.

2.2 Collection of expired air and skin gases

The expired air was collected in Douglas bag, and skin gas was collected in the sampling bag at rest, during and after cycle and hand-grip exercises. In skin gas sampling, the right hand was cleaned, and wiped with paper before skin gas collection. After the hand was inserted into the sampling bag, which was fixed to the elbow (cycle exercise) or the middle of the upper arm (the center point between olecranon and acromion) (Fig 1-A; cycle exercise; Fig 1-B; hand-grip exercise). All the air in the bag was sucked out with a glass syringe; The gas in the bag was replaced with 600ml nitrogen gas. The skin gas collected for 1 min, and measured acetone concentration by gas chromatography.

2.3 Analytical procedure and conditions

The sampling gas was analyzed with a cold trap gas chromatographic system (Yamane et al. 2006; Nose et al. 2006). The sample was automatically sucked into the stainless-steel tube from loop cooled with liquid nitrogen. After the sample injection valve was rotated from the trapping position to the injection position, the trap tubing was heated directly to aid thermal desorption of the acetones in the sample. Acetone concentration in gas samples was detected by a gas chromatography with a flame ionization (FID).

The concentration of acetone in expired air and skin gases was measured using modified methods of Yamane et al. (2006). Acetone was detected with a flame ionization detector

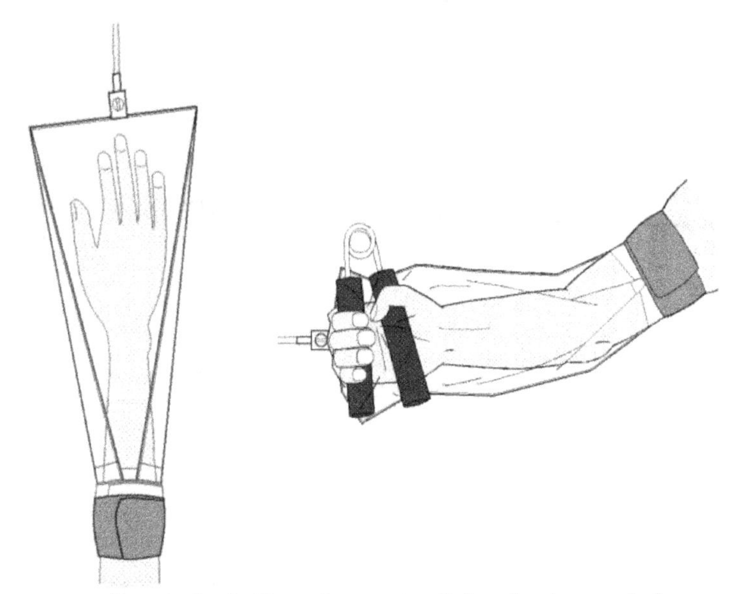

Fig. 1. Skin gas sampling methods (A: cycle exercise, B: hand-grip exercise).

(FID) by a Type GC-14B gas chromatograph (Shimadzu, Kyoto, Japan). The separation conditions for measuring the acetone were as follows: a Porapack Q capillary column (Type G-950: 1.2 mm internal diameter and 5.8 m long, Chemical Evaluation and Research Institute, Tokyo, Japan); the injection and detection temperature was 150℃; the column temperature was 100℃; and the retention time was 2.3 min.

3. Results

3.1 Hand-grip exercise

Table 1 demonstrated the mean (M) and standard deviation (S.D.) of skin gas acetone concentration at rest, during exercise, after exercise, and each coefficient of variation (C.V.) when one subject performed the same exercise four times. The C.V. of at rest, during exercise and recovery were 10.98, 9.52, 9.73, 8.67 and 8.49, respectively. Fig.2 demonstrates changes in acetone emanated from skin gas when one subject performed the same hand-grip exercise four times. The skin gas acetone concentration significantly increased during exercise compared to resting level, and decreased to almost the same as basal level immediately after 1.0-2.0 min during recovery.

	Rest	Exercise	Recovery		
			1.0-2.0 min	3.0-4.0 min	5.0-6.0 min
Mean (ppm)	0.082	0.147	0.113	0.105	0.106
SD	0.009	0.014	0.011	0.0091	0.009
C.V. (%)	10.98	9.52	9.73	8.67	8.49

Table 1. Mean, SD and CV of skin gas acetone concentration at rest, during exercise and recovery.

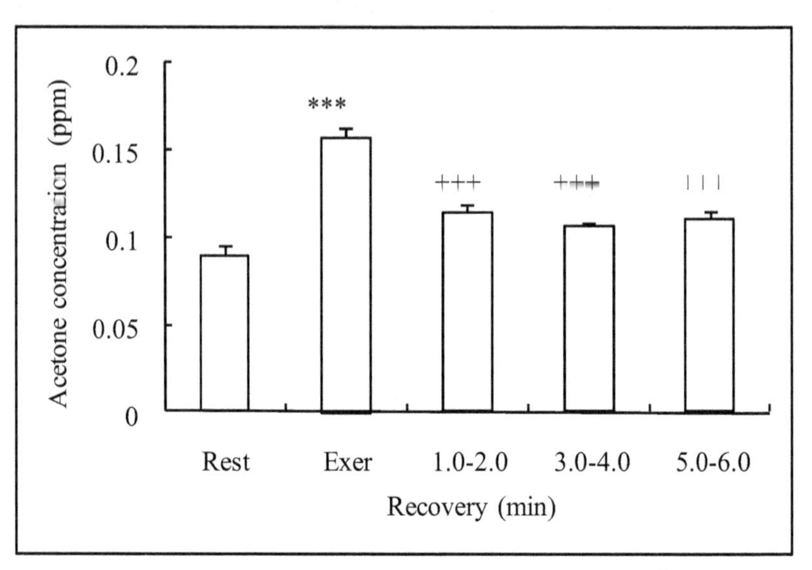

Fig. 2. Skin gas acetone concentration at rest, during hand-grip exercise and recovery. One subject performed the same hand-grip exercise four times. Values are M±SEM. **p<0.01, significant difference compared to rest ; ++p<0.01 significant difference compared to exercise.

Acetone concentration in skin gas during hand-grip exercise 2 (30kg ×20 times) was significantly higher than basal level (Figure 3). Although skin gas acetone during exercise 1 (20kg×30 times) and exercise 3 (10kg×60 times) increased, significant difference was not found. No significant difference was found in skin gas acetone concentration during hand-grip exercise among exercise 1, 2, and 3.

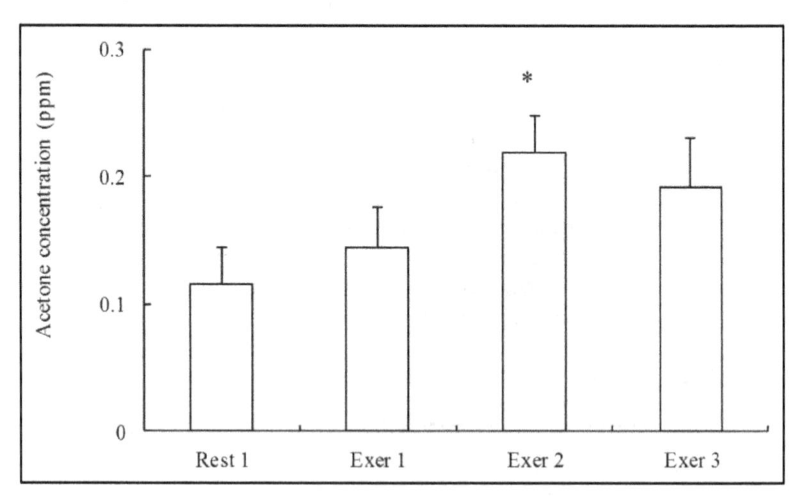

Fig. 3. Comparison of skin-gas acetone concentration among exercise 1, 2 and 3. Values are M±SEM. *p<0.05, significant difference compared to rest.

3.2 Cycle exercise

Figure 4 represents the changes in acetone concentration in expired air at the basal level, 360, 720 and 990 kgm/min during cycle exercise and recovery of 5 min after the 990 kgm/min exercise. Acetone concentration of expired air tend to increase with an increase in exercise intensity. The acetone concentration of expired air at 990 kgm/min significantly increased compared with the basal level ($p<0.05$). Figure 5 shows the changes in acetone excretion in expired air at the basal level, during cycle exercise and recovery. The acetone

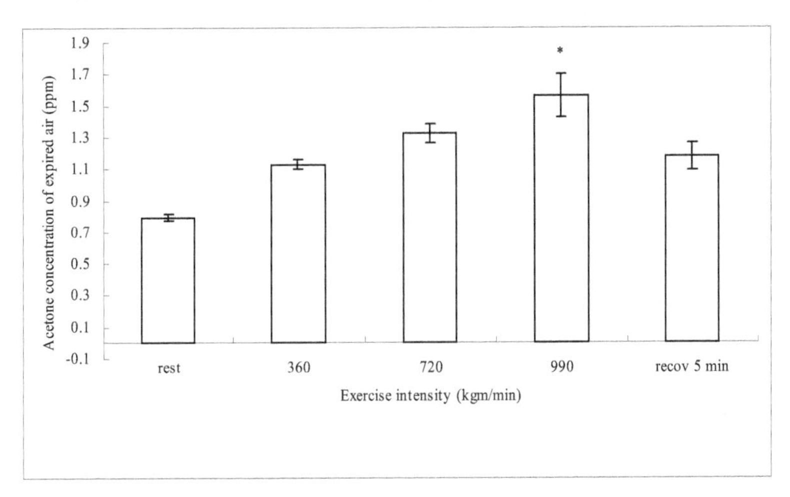

Fig. 4. Changes in acetone concentration in expired air at the basal level, during cycle exercise and recovery.*$p<0.05$ significant difference compared to rest.

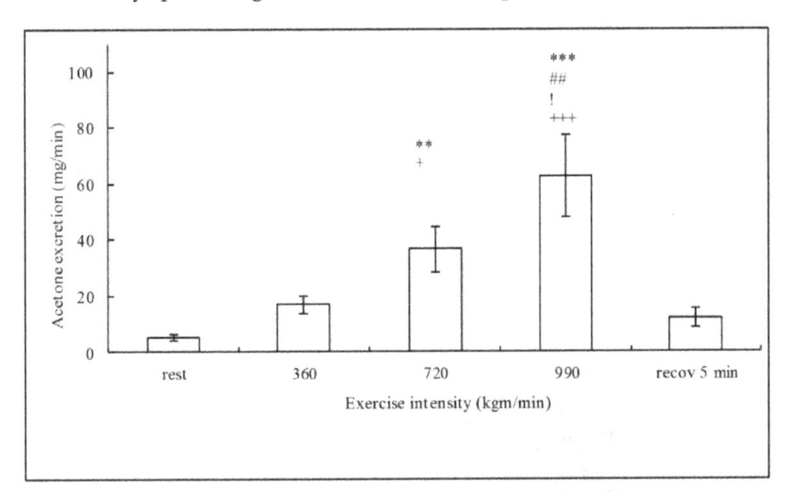

Fig. 5. Changes in acetone excretion in expired air at the basal level, during exercise and recovery. *** $p<0.001$, **$p<0.01$ significant difference compared with resting value. ## $p<0.01$ significant difference compared to 360 kgm/min, ! $p<0.05$ significant difference compared to 720 kgm/min, +++ $p<0.001$, + $p<0.05$ significant difference compared to 5 min after exercise.

excretion at 720 and 990 kgm/min significantly increased compared with the basal level ($p<0.05$). As shown in Figure 6, the skin gas acetone concentration at 990 kgm/min significantly increased compared with the basal level and 360 kgm/min ($p<0.05$). Acetone concentration in expired air was 4-fold greater than skin gas at rest and 3-fold greater during exercise ($p<0.01$). There is a significant relationship between skin gas acetone concentration with expired air ($r=0.752$, $p<0.01$, Fig 7)

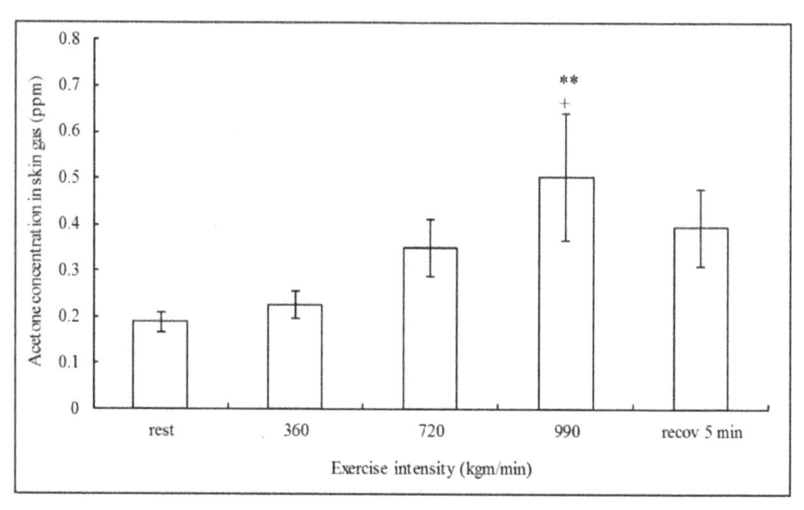

Fig. 6. Changes in skin gas acetone concentration at the basal level, during exercise and recovery. ** $p<0.01$ significant difference compared to rest. # $p<0.05$ significant difference compared to 360kgm/min.

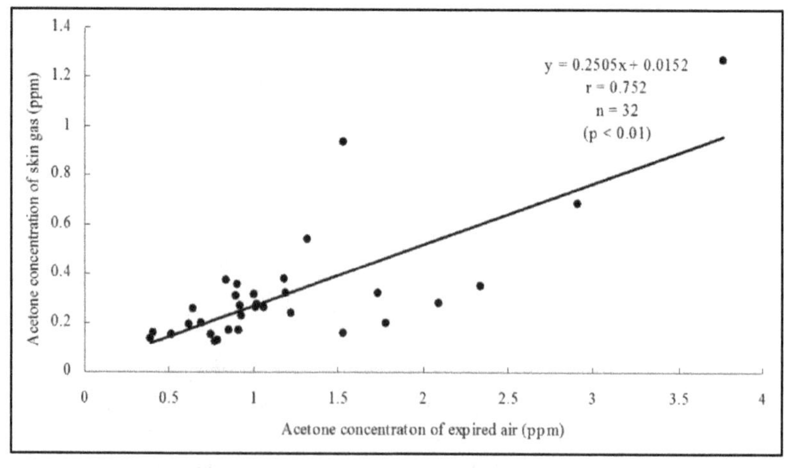

Fig. 7. The relationship in acetone concentration between in expired air and skin gas.

4. Discussion

Acetone is drived from spontaneous, non-enzymatic, and decarboxylation of acetoacetate when there is an insufficient supply of blood glucose (Owen et al. 1982). The relationship between acetone concentration in blood and breath has been well established (Owen et al. 1982). Previously, it has been demonstrated that skin gas acetone is correlated with breath acetone (Naitoh et al. 2002). Yamane et al (2006) also reported that skin acetone concentrations of patients with diabetes were significantly higher than those of the control subject. Naito al. (2002) reported that concentrations of acetone in both skin gas and breath increased according to the length of fasting periods, and also demonstrated that there is a good relationship between skin and breath acetone concentration.

It is a well known fact that ketogenesis is enhanced in diabetes, fasting and exercise (Féry et al. 1983; Balasse and Féry (1989); Wahren et al. 1984). Balasse and Féry (1989) reported that plasma ketone bodies increased after treadmill running at the intensity of 40% ~ 60% maximal aerobic capacity for 120 min compared with before exercise, and also observed that the plasma ketone bodies was higher in 60% maximal power compared to that of 40 % maximal power.

The C.V. of skin gas acetone concentration during hand-grip exercise was 9.52, when skin gas acetone measured four times in one subject. These results indicate that reasonable reproducibility is obtained in measurements of skin gas acetone concentration during hand-grip exercise and that determination of skin gas acetone is highly reliable.

The Exercise 2 of the hand-grip exercise (30kg ×20 times) showed an increase in acetone level of skin gas compared to basal level. The exercise 1 (20kg×30 times) and 3 (10kg×60 times) tend to increase skin gas acetone, but the increase was not significant. Although the total work is the same between Exercise 1,2 and 3, acetone concentration in Exercise 2 was increased compared to rest. The results of our study coincidently is comparable with the result from previous study (Balasse and Féry 1989). These experiments suggest that acetone concentration in blood and skin gas are related to intensity of exercise.

Senthilmohan et al. (2000) have demonstrated that acetone concentration of expired air increased in running exercise (Senthilmohan et al. 2000), but was not clear whether the acetone concentrations of skin gas increased with an elevation of exercise intensity. Acetone concentration of expired air and skin gas at the intensity of 990kgm/min has significantly increased compared to the basal level (Fig. 5 and 6). These results basically agreed with those of other studies (Balasse and Féry 1989; Féry et al.1983; Féry et al.1986). Sasaki et al. (2011) demonstrated in the graded exercise test that acetone level in exhaled air began to increase at the intensity of 39.6 % of maximal oxygen uptake. From these results, it is clear that acetone concentration of expired and skin gases increased with elevation of exercise intensity.

FFA and ketone bodies in blood increase during prolonged exercise and increase in mobilization results in elevation of utilization (Loy et al. 1986; Ravussin et al. 1986; Sasaki et al. 2011). The acetone concentration correlates with the concentration of blood ß-hydroxybutyrate (Reichard et al. 1979; Yamane et al. 2006). The kinetic responses of ketone bodies to exercise is very complex. The increase in ketogenesis is related to the rise in FFA levels and to an increase in the ketogenic capacity of the liver (Balasse and Féry 1989). Exercise is known to increase the hepatic conversion of FFA to ketones in control group (Wahren et al. 1984). However, Wahren et al. (1975) have shown that uptake of ketone bodies by the leg muscle was of minor importance during bicycle exercise in control subject. Wahren et al. (1984) suggested that the rate of hepatic ketogenesis can be influenced by the rate of lipolysis (FFA release from adipose tissue) and the blood flow to the liver, and by the fractional extraction of FFA by the liver and an increase in the proportion of FFA converted to ketones. Furthermore, the increase of plasma catecholamines in response to the exercise may have contributed to the augmented fractional extraction of FFA by the splanchnic bed, which was observed in the control in response to exercise (Wahren et al. 1984). Dynamic hand-grip exercise increased norepinephrine (Costa et al. 2001). The increase of skin gas acetone might have been caused by norepinephrine secretion, which was increased during dynamic hand-grip exercise.

5. Acknowledgement

It was concluded in this report that acetone concentration in expired air and skin gas is directly related to exercise intensity. This human skin gas project was supported by the Aichi Science and Technology Foundation.

6. References

Balasse EO, Féry F. Keton body production and disposal: effects of fasting, diabetes, and exercise. Diabetes Metab Rev 1989; 5:247-270.

Costa F, Christensen NJ, Farley G, Biaggioni I. NO modulates norepinephrine release in human skeletal muscle: implications for neural preconditioning. Am J Physiol. Regul Integr Comp Physiol 2001; 280: R1494-R1498.

Féry F, Blasse EO. Ketone body turnover during and after exercise in overnight-fasted and starved humans. Am J Physiol 1983; 245: E318-E325.

Féry F, Blasse EO. Response of ketone body metabolism to exercise during transition from postabsorptive to fasted state. Am J Physiol 1986; 250: E495-E501.

Laffel L. Ketone bodies: a review of physiology, pathophysiology and application of monitoring to diabetes. Diabetes Metab Res Rev 1999; 15: 412-426.

Loy SF, Conlee RK, Winder WW, Nelson AG, Arnall DA, Fisher AG: Effect of 24-h fast on cycling endurance time at two different intensities. J Appl Physiol 1986; 61: 654-659.

Miekisch W, Schubert JK, Noeldge-Schomburg GFE. Diagnostic potential of breath analysis-focus on volatile organic compounds. Clinica Chimica Acta 2004; 347: 25-39.

Mitchell GA, Kassovska-Bratinova S, Boukaftane Y, Robert MF, Wang SP, Ashmarina L, Lambert M, Lapierre P, Potier E. Medical aspects of ketone body metabolism. Clin Invest Med 1995; 18: 193-216.

Mori K, Funada T, Kikuchi M, Ohkuwa T, Itoh H, Yamazaki Y. Tsuda T. Influence of dynamic hand-grip exercise on acetone in gas emanating from human skin. Redox Rep 2008; 13:139-142.

Musa-Veloso K, Likhodii SS, Cunnane SC. Breath acetone is a reliable indicator of ketosis in adults consuming ketogenic meals. Am Clin Nutr 2002; 76: 65-70.

Naitoh K, Inai Y, Hirabayashi T, Tsuda T. Direct temperature-controlled trapping system and its use for the gas chromatographic determination of organic vapor released from human skin. Anal Chem 2000; 72: 2797-2801.

Naitoh K, Tsuda T, Nose K, Kondo T, Takasu A, Hirabayashi T. New measurement of hydrogen gas and acetone vapor in gases emanating from human skin. Instrum Sci Technol. 2002; 3: 267-280.

Nose K, Ueda H, Ohkuwa T, Kondo T, Araki S, Ohtani H, Tsuda T. Identification and assessment of carbon monoxide in gas emanated from human skin Chromatography 2006; 27: 63-65.

Owen OE, Trapp VE, Skutches CL, Mozzoli MA, Hoeldtke RD, Boden G, Reichard GA. Acetone metabolism during diabetic ketoacidosis. Diabetes. 1982; 31: 242-248.

Ravussin E, Bogardus C, Scheidegger K, Lagrange B, Horton ED, Horton ES. Effect of elevated FFA on carbohydrate and lipid oxidation during prolonged exercise in humans. J Appl Physiol 1986; 60: 893-900.

Reichard GA, Haff AC, Skutches CL, Paul P, Holroyde CP, Owen OE. Plasma acetone metabolism in the fasting human. J Clin Invest 1979; 63: 619-626.

Sasaki H, Ishikawa S, Ueda H, Kimura Y. Acetone response during graded and prolonged exercise. Yoshikawa T, Naito Y (eds): Gas Biology Research in Clinical Practice. Basel, Karger, 2011,pp119-124.

Schubert JK, Miekisch W, Geiger K, Noldge-Schomburg GFE. Breath analysis in critically ill patients: potential and limitations. Expert Rev Mol Diagn 2004; 4: 619-629.

Senthilmohan ST, Milligan DB, McEwan MJ, Freeman CG, Wilson PF. Quantitative analysis of trace gases of breath during exercise using the new SIFT-MS technique. Redox Rep 2000; 5: 151-153.

Tassopoulos CN, Barnett D, Fraser TR. Breath-acetone and blood-sugar measurements in diabetes. Lancet 1969; 28: 1282-1286.

Tsuda T, Ohkuwa T, Itoh H. Findings of skin gases and their possibilities in healthcare monitoring. Yoshikawa T, Naitoh Y (eds): Gas Biology Research in Clinical Practice. Basel, Karger, 2011,pp125-132.

Turner C, Parekh B, Walton C, Spanel P, Smith D, Evans M. An exploratory comparative study of volatile compounds in exhaled breath and emitted by skin using selected ion flow tube mass spectrometry. Rapid Commun Mass Spectrom 2008; 22: 526-532.

Wahren J, Hagenfeldt L, Fehlig P. Splanchnic and leg exchange of glucose, amino acids, and free fatty acids during exercise in diabetes mellitus. *J Clin Invest* 1975; 55: 1303-1314.

Wahren J., Sato Y., Ostman J., Hagenfeldt L., Felig P. Turnover and splanchnic metabolism of free fatty acids and ketones in insulin-dependent diabetics at rest and in response to exercise. J Clin Invest 1984; 73: 1367-1376.

Yamai K, Ohkuwa T, Ito H, Yamazaki Y, Tsuda T. Influence of cycle exercise on acetone in expired air and skin gas. Redox Rep. 2009; 14: 285-289.

Yamane N, Tsuda T, Nose K, Yamamoto A, Ishiguro H, Kondo T. Relationship between skin acetone and blood β-hydroxybutyrate concentrations in diabetes. Clin Chimi Acta 2006; 365: 325-329.

9

Gas Chromatography in the Analysis of Compounds Released from Wood into Wine

Maria João B. Cabrita[1], Raquel Garcia[1], Nuno Martins[1], Marco D.R. Gomes da Silva[2] and Ana M. Costa Freitas[1]
[1]*School of Science and Technology, Department of Plant Science, Institute of Agricultural and Environmental Science ICAAM Mediterranean, University of Évora, Évora,*
[2]*REQUIMTE, Department of Chemistry, Faculty of Science and Technology, New University of Lisbon, Campus Caparica,*
Portugal

1. Introduction

Wood has been used in alcoholic beverages for centuries, mainly as material for containers used for alcoholic beverages aging. Recently OIV (Organisation International de la Vigne et du Vin) approved the use of chips (Resolution oeno 3/2005) and staves as alternatives for barrels. These practices are being rapidly spread among winemakers. The increased used of these alternatives are mainly related to low investments, similar sensorial results obtained in shorter time, simplicity of use and the possibility of avoiding contamination and off-flavours, too-often related to aged or contaminated barrels.

Besides oak, other woods are being looked at for enological purposes, such as acacia, cherry, chestnut and mulberry. Their characteristics are commonly compared to oak. In the past, chestnut (*Castanea sativa*) was widely used in the Mediterranean area, because of its availability and its cheap price. Chestnut wood has higher porosity than oak. Cherry wood (*Prunus avium*) has high porosity and oxygen permeation, and is usually used for short aging times. Acacia wood (*Robinia pseudoacacia*) is hard, with low porosity. Mulberry wood (*Morus alba* and *Morus nigra*) is tender and elastic, with medium porosity, and is characterized by a low release of compounds. The lack of properties for cooperage is now overcome by their possible use as staves or chips.

The aim of this work is to present an overview on volatile and semi-volatile composition of different kind of wood with oenological interest. Within this purpose, this work will be focused on a bibliographic review of the most used chromatographic methods for characterization of volatile and semi-volatile compounds, including also a brief description of the most common reported sample preparation methods for chromatographic analysis.

The composition of woods volatile fraction depends on the botanical species and geographic origin. Prior to its final use a toasting step is needed. This toasting process induces the formation of a great number of volatile and odoriferous compounds, and it is the main reason for significant differences among non-toasted and toasted wood. Volatiles and semivolatiles compounds from wood belong to several different chemical families: lactones, terpenes, norisoprenoids, aldehydes, ketones, alcohols, phenols and esters.

Because of the odourific impact of some volatiles, a number of studies dealt with their characterization in wood-aged beverages and woods itself. Volatile phenols such as guaiacol and eugenol, phenolic aldehydes such as vanillin and syringaldehyde, oak lactones, and furanic aldehydes (furfural and derivatives) have been described as the main contributors to the sensory fingerprinting of aged alcoholic beverages.

Most of the volatile compounds are formed during open-air seasoning and toasting phases of wood processing. Several furanic aldehydes and ketones come from the thermo degradation of celluloses and hemicelluloses. This is the case of hydroxy-methylfurfural (HMF) (from cellulose-derived glucose) and furfural arising from pentoses produced by partial hydrolysis of hemicelluloses. These latter compounds are responsible for almond and toasty odours. Thermal degradation of lignin determines the formation of methoxylated volatile phenols (*i.e.*, guaiacyl and syringyl derivatives), phenolic ketones, and phenolic aldehydes contributing to smoked or spiced and vanilla aromas, respectively.

To study volatiles in woods or in alcoholic beverages some preparation methods must be considered. The volatile fraction from woods can be studied from three different points of view: the volatiles existing in wood itself, the volatiles that wood can release into synthetic medium miming alcoholic beverages and the study of alcoholic beverages like wines and spirits after aging with wood contact. In practical terms, that implies using sample preparation methods for solid and liquid matrices.

Sample preparation for gas chromatography (GC) analysis of volatile compounds from woods can be performed by pressurized liquid extraction, a process that combines temperature and pressure with liquid solvent to achieve extracts rapidly and efficiently. This technique is available in an automated or a manual version as accelerated solvent extraction (ASE), and it is being currently used as a good alternative for the solid-liquid extraction used previously for studying volatiles from solid matrices. Most of the "oldest" sample preparation methods involved Soxhlet extraction with dichloromethane or methanol or simultaneous distillation-extraction procedures.

The sample preparation procedures for GC analysis of volatile compounds in wines or hydro-alcoholic mixtures in general, are usually done by LLE (liquid-liquid extraction), SPE (solid-phase extraction) or SPME (solid-phase microextraction). But, more environmental friendly sample preparation techniques like SBSE (stir bar sorptive extraction) can also be employed with good results when thermal desorption is used instead of retro-extraction, for subsequent chromatographic analysis.

The chromatographic technique more suitable for these analyses is gas chromatography-mass spectrometry (GC/MS), as the mass spectrometer allows a more powerful tool for compounds identification. When MS is coupled with two-dimensional (2D) chromatographic systems, broader capabilities can be open in order to fully characterized wine matrices. An olfactometric approach is also mandatory in order to identify chromatogram zones where peak identification should take place. Consequently gas chromatography-olfactometry (GC-O) emerges as an important technique to achieve this purpose. GC-O analysis has been widely used to identify odour-active compounds or to screen the odour volatile composition in wines. This technique allows to obtain relatively simple olfactograms and to establish a hierarchy of the most important odorants according to their potential sensory impact. In general it implies the use of headspace technique as sample preparation method, such as purge-and-trap (P&T) or solid-phase microextraction (HS-SPME) systems.

2. Composition and biosynthesis of the volatile fraction of woods: Effect of botanical species and toasting

For centuries, wood has been used in wine technology, either as containers for transportation either for aging wine and spirits. Over the years, many different tree species have been used in cooperage, but oak and chestnut woods proved to be the best ones to manufacture barrels and therefore the most suitable ones for wine and spirits aging.

The use of barrels is, however, expensive since barrels can only be used few times for wine aging. Besides the above mentioned issue, barrels can often be the cause of wine contaminations and off-flavours, if sanitation procedures are not correctly applied.

All woods are composed of cellulose, lignin, ash-forming minerals, and extractives formed into a cellular structure. The characteristics and amounts of these compounds and differences in cellular structures give to each wood its specific characteristics. Some are more flexible than others, some are easier to work, some are harder or softer, heavier or lighter.

Extractives from different wood species comprise several substances which belong to an extremely wide range of chemical families that are characteristics of each wood. Thus, extractives allow the identification of wood species, contributing also for the determination of other wood properties, namely permeability, density, hardness and compressibility.

Besides botanical species, geographical origin also contributes to the final chemical composition of the extractives. Technological treatments during cooperage such as seasoning and toasting give the final characteristics to the barrel (Chatonnet & Dubourdieu, 1998; Cadahía et al., 2001; Doussot et al., 2002; Pérez-Prieto et al., 2002; Cadahía et al., 2003).

Wood physical properties play a significant role in cooperage. Usually, oak is the most used specie due to its unique physical and chemical nature. Physically, oak has strength since its wide radial rays give strength when shaped for a cask. Oak is also a "pure wood" as opposed to pine or rubber trees which contain resin canals that can transfer strong flavours to maturing beverages.

Several studies have been conducted in order to evaluate the accumulation of oak volatile compounds in wines and spirits and how different factors affect the final concentration of these compounds both in barrels (Puech, 1987; Sefton et al., 1993; Yokotsuka et al., 1994; Waterhouse & Towey, 1994; Piggot et al., 1995; Escalona et al., 2002; Ferreras et al., 2002; Pérez-Prieto et al., 2002, 2003; Netto et al., 2003; Madrera et al., 2003; Garde-Cerdán et al., 2002, 2004; Gómez-Plaza et al., 2004; Jarauta et al., 2005; Garde Cerdán & Ancín-Azpilicueta, 2006 b; Caldeira et al., 2006 a,b; Jiménez-Moreno & Ancín-Azpilicueta, 2007; Jiménez-Moreno et al., 2007; Frangipane et al., 2007) and with oak chips (Pérez-Coello et al., 2000; Arapitsas et al., 2004; Caldeira et al., 2004; Guchu et al., 2006; Frangipane et al., 2007; Bautista-Ortin et al., 2008; Rodríguez-Bencomo et al., 2008, 2009; Rodríguez- Rodríguez & Gómez-Plaza, 2011).

Other studies were made in model solutions to avoid matrix influence (Singleton & Draper, 1961; Ancín-Azpilicueta et al., 2004; Fernández de Simón et al., 2010 a; Rodríguez-Rodríguez & Gómez-Plaza, 2011) or in wood itself (Vichi et al., 2007; Natali et al., 2006; Díaz-Maroto et al., 2008).

More recently some studies have been made with other woods than oak (Flamini, et al., 2007; de Rosso et al., 2009; Fernández de Simón et al., 2009; Caldeira et al., 2010) as they becoming increaseling interesting.

Ever since OIV approved the use of alternatives for barrels, like chips and staves, researchers are looking into other kind of woods such as acacia, cherry and mulberry, which were abandoned in the past due to the lack of cooperage properties.

2.1 Effect of botanical species

The main botanical species used in cooperage is oak (genus *Quercus*) and in a lesser extension, chestnut (genus *Castanea*). The oak species more used in cooperage are *Quercus alba* (American oak) growing in different areas of the United States, *Quercus patraea* Liebl. (sessili oak), *Quercus Robur* L. (pedunculate oak) and *Quercus pyrenaica* growing in Europe. French oaks are the most widespread wood for barrels, especially those from Allier and Limousin regions. But oaks from Hungary, Russia, Spain, Romania and Portugal are also becoming more and more attractive.

Many researchers are now dedicated to the study of chemical compounds in different kind of woods, although oak is still the most studied one. Whenever a new wood is tested it is compared with the oak effects in wines and spirits. Oak belonging to different species can be significantly different regarding volatile composition. Jordão et al. (2006) concluded that *Quercus pyrenaica* released significantly more volatile compounds than *Quercus patraea*, in a study with SPME in hydro-alcoholic oak wood solutions. Fernández de Simón et al., (2010 a) concluded that *Quercus pyrenaica* chips, althought similar to other species, have some odourific particularities, such as high levels of furanic compounds, eugenol, furaneol and *cis*-whiskylactone and low levels of vanillin. Several authors pointed out that β-methyl-γ-octalactone (whisky lactone), particularly the *cis* isomer, can be used to differentiate American from French oaks (Masson et al., 1995; Waterhouse & Towey, 1994; Masson et al., 1997; Chatonnet & Dubourdieu, 1998; Pérez-Coello et al, 1999). Several norisoprenoid compounds were found in American oak, but were almost absent in European oaks (Sefton et al., 1990). Same wood species from French oak and East European oaks can be distinguished by their contents of eugenol, 2-phenylethanol and aromatic aldehydes, mostly vanillin and syringaldehyde (Prida & Puech 2006).

The chemical compounds released from acacia, chestnut, cherry, mulberry and oak untoasted woods into alcoholic extracts, shows that each wood has a different and characteristic profile (Flamini et al., 2007, de Rosso et al., 2009, Fernández de Simon et al., 2009). Acacia had significant aromatic aldehydes, particularly vanillin, syringaldehyde and dihydroxy-benzaldehyde but no eugenol (de Rosso et al., 2009) or methoxyeugenol (Flamini et al., 2007) were found. Chestnut and oak showed the highest content of volatile compounds namely vanillin, eugenol, methoxyeugenol, syringaldehyde, α-terpineol, and oak presented high amouts of *cis*- and *trans*-β-methyl-γ-octalactones. In cherry, several aromatic compounds were found although in low abundance, but mulberry was the poorest, with small amounts of eugenol and absence of methoxyeugenol (de Rosso et al., 2009). Cherry is also characterized by methoxyphenols, particularly, high content of trimethoxyphenol (Flamini et al., 2007). Chestnut and oak woods are known to release significant amounts of eugenol and methoxyeugenol into wines. High amounts of vanillin are released from chestnut and high levels of syringaldehyde are released from acacia and oak woods.

2.2 Effect of toasting level

Oak wood chemical composition mainly depends on the species, its provenience and the various treatments that wood undergoes in cooperage, such as seasoning, region of seasoning and toasting (Marco et al., 1994; Chatonnet & Dubourdieu, 1998; Cadahía et al., 2001; Doussot et al., 2002; Pérez-Prieto et al., 2002; Cadahía et al., 2003).

Seasoning prevents the wood from shrinking after barrel production while firing is applied to stabilise the curved shape of the barrel (van Jaarsveld et al., 2009) and both these steps play a crucial role in wood flavour development.

Wood seasoning in cooperage is usually performed under natural conditions in open air during a variable period of time (18 to 36 months), but artificial seasoning can also be performed. However, natural seasoning has a more positive effect on the odourific profile of wood. During this process the wood volatile compounds profiles, which include lactones, phenolic aldehydes or volatile phenols, show significant differences.

Nevertheless, among all variables that can influence the impact of woods in wine or spirits sensory changes, heating is the most important. Toasting has a significant influence on wood's chemical compounds, modifying both, the quantity and the quality of the extractable substances (Cutzach et al., 1997; Chatonnet et al., 1999). The toasting process drastically enhances the gain in volatile compounds arising from the thermal degradation of oak wood (Cutzach et al., 1997; Chatonnet et al., 1999; Doussot et al., 2002).

When heat is applied to wood during toasting process, chemical bonds are disrupted within biopolymers such as cellulose, hemicellulose, lignin, polysaccharides, polyphenols and lipids, resulting in degradation or compositional changes by pyrolysis and thermolysis (Fernández de Simón et al., 2009; van Jaarsveld et al., 2009), which induce a notable modification of wood chemical composition.

Volatile phenols, phenolic aldehydes, phenyl ketones and some phenyl alcohols are mainly formed from lignin thermodegradation. In particular, high levels of mono and dimethoxylated phenols, benzoic and cinnamic aldehydes were identified in toasted wood.

Heat degradation of polyosides leads to the production of furanic aldehydes, pentacyclic and hexacyclic ketones. Lactones, formed from wood lipids, increase their concentration at the beginning of toasting, although they can be destroyed by a lengthy toasting process (Giménez-Martínez et al., 1996; Chatonnet et al., 1999; Cadahía et al., 2003). This thermo degradation process leads to the formation of several compounds that can be transferred to wine and spirits during aging or maturation, hence, extensive information about the volatile composition of wood used in cooperage would be of great interest to the wine industry. Table 1 resumes information regarding main volatile compounds formed from wood biopolymer during toasting process.

Common name	IUPAC name	m/z fragment	LRI Apolar column	LRI Polar column	Aroma notes
DERIVED FROM POLYSACHARIDES					
Furanic aldeydes					
Furfural	2-Furancarboxaldehyde	95,**96**	834(a) 811(h)	1444(b)	Bread, sweet almond (i), slightly toasty, caramel(h)
5-Methylfurfural	5-Methyl-2-furancarboxaldehyde	109,**110**	965(a)	1551(b)	Almond, caramel(i), spicy, toasty(h)
5-Hydroxy methylfurfural	5-Hydroxymethyl-2-furancarboxaldehyde	**97**,126	1235(a)	2466(b)	Odourless(n)
Maltol	3-Hydroxy-2-methyl-4*H*-pyran-4-one	71,**126**	1111(a)	1938(b)	Toasty(n)
Furaneol	4-Hydroxy-2,5-dimethyl-3(2*H*)-furanone	85,**128**	1083(a)	2013(b)	Caramel-like(o)
Acids					
Acetic acid	Ethanoic acid	**43**,45,60	602(g)	1464(f)	vinegar(f)
DERIVED FROM LIGNIN AND POLYPHENOLS					
Volatile phenols					
Eugenol	2-Methoxy-4-(prop-2-enyl) phenol	77,103, **164**	1359(a)	2139(b)	Clove,honey (i) spicy, cinnamon(h)
Isoeugenol (*cis* and *trans*)	*cis-* and *trans*-1-Methoxy-4-(prop-2-enyl) phenol	77,149, **164**	1408(a) 1451(a)	2226(b) 2314(b)	Spicy (m) clove,woody(h)
Phenol	Hydroxybenzene	66,94	983(a)	1978(b)	Ink(n)
3,4-Dimethylphenol	1-Hydroxy-3,4-dimethylbenzene	**107**	1193(a)	2192(b)	
o-Cresol	2-Methylphenol	**107**/108	1059(a)	1980(b)	Leather, spicy(m)
p-Cresol	4-Methylphenol	**107**/108	1079(a)	2056(b)	
m-Cresol	3-Methylphenol	**107**/108	1086(a)	2064(b)	
Guaiacol	2-Methoxyphenol	109/**124**	1089(a)	1833(b)	Smoke,sweet, medicine (i)
4-Methylguaiacol	4-Methyl-2-methoxyphenol	123/**138**	1191(a)	1928(b)	Spicy,phenolic, lightly green(h)
4-Ethylguaiacol	4-Ethyl-2-methoxyphenol	**137**/152	1274(a)	2002(b)	Phenolic(i,m) leather(m) smoked(i)
4-Propylguaiacol	4-Propyl-2-methoxyphenol	**137**/166	1461(a)	2083(b)	Leather animal(m)

Common name	IUPAC name	m/z fragment	LRI Apolar column	LRI Polar column	Aroma notes
4-Vinylguaiacol	4-Vinyl-2-metoxyphenol	135/**150**	1314(a)	2165(b)	Clove (i)
Syringol	2,6-Dimethoxyphenol	139/**154**	1353(a)	2237(b)	Smoke, burned wood(f)
4-Methylsyringol	4-Methyl-2,6-dimetoxyphenol	168	1449(a)	2322(b)	Smoke,burned, flowery(f)
4-Ethylsyringol	4-Ethyl-2,6-dimethoxyphenol	167	1528(a)	2381(b)	
4-Allylsyringol	4-Allyl-2,6-dimethoxyphenol	194	1605(a)	2511(b)	Spicy smoky(m)
4-Propylsyringol	4-Propyl-2,6-dimethoxyphenol	167	1612(a)	2452(b)	
4-Ethylphenol	1-Ethyl-4-hydroxybenzene	**107**,122		2201(f)	Animal, horse, stable(f)
2-Phenylethanol	Hydroxyethylbenzene	91	1113(a)	1888(b)	Floral, roses(h)
Phenolic aldeydes					
Vanillin	4-Hydroxy-3-methoxybenzaldehyde	**151**/152	1399(a)	2518(b)	Vanilla (i)
Benzaldehyde	Phenylmethanal	77,**106**	962(a)	1493(b)	Bitter almonds (h)
Conyferaldehyde	3-Methoxy-4-hydroxycinnamaldehyde	178	1747(a)	3096(b)	Vanilla, woody(n)
Syringaldehyde	4-Hydroxy-3,5-dimethoxybenzaldehyde	181,**182**	1643(a)	2904(b)	Vanilla (i)
Sinapaldehyde	3,5-Dimethoxy-4-hydroxycinnamaldehyde	208	2002(a)	3458(b)	Vanilla(n)
Phenolic esters					
Ethyl vanillate	4-Hydroxy-3-methoxy benzoic acid, ethyl ester	**151**, 167,196	1648(g)		Flower, vanilla,fruit, sweet, (i)
Methyl vanillate	4-Hydroxy-3-methoxy benzoic acid, methyl ester	**151**,182	1518(a)	2565(b)	Caramel, butterscotch, vanilla (i)
Phenyl ketones					
Acetovanillone	1-(4-Hydroxy-3,5-dimethoxyphenyl)-ethanone	**151**/166	1487(a)	2595(b)	Vanilla(i)
Propiovanillone	1-(4-Hydroxy-3,5-dimethoxyphenyl)-propanone	**151**/180	1501(a)	2661(b)	Vanilla(n)
Butyrovanillone	1-(4-Hydroxy-3-methoxyphenyl)-butanone	**151**/194	1590(a)	2770(b)	Caramel, sweet, Buttery (i)

Common name	IUPAC name	m/z fragment	LRI Apolar column	LRI Polar column	Aroma notes
Acetosyringone	2-(4-Hydroxy-3-methoxyphenyl)-acetaldehyde	**181**/196	1744(a)	2953(b)	
Propiosyringone	1-(4-Hydroxy-3,5-dimethoxyphenyl-) propanone	**181**/210	1753(a)	3010(b)	Vanilla(n)
Isoacetosyringone	2-(4-Hydroxy-3,5-dimethoxyphenyl)-acetaldehyde	167	1712(a)	2927(b)	
Isopropiosyringone	1-(4-Hydroxy-3,5-dimethoxyphenyl)-2-propanone	167	1785(a)	2979(b)	
Alcohols					
Coniferyl alcohol	4-(1-*trans*)-3-Hydroxy-prop-1-enyl-2-methoxyphenol	137	1745(a)	3213(b)	
Benzyl alcohol	Hydroxy methylbenzene	79,108		1879(p)	Sweet, floral (m)
DERIVED FROM LIPIDS					
Lactones					
β-Methyl-γ-octalactone (*cis*)	*cis*-4-Methyl-5-butyldihydro-2(3*H*)-furanone	99	1325(a)	1928(b)	Sweet, coconut (j) woody (k,l)
β-Methyl-γ-octalactone (*trans*)	*trans*-4-Methyl-5-butyldihydro-2(3*H*)-furanone	99	1292(a)	1861(b)	Sweet, coconut (j) woody (k,l)
γ-Butyrolactone	Dihydro-2(3*H*)-furanone	86	913(a)	1593(b)	Cheese (m)
Acids					
Propionic acid	Propanoic acid	45,**74**	668(g)		Fruity, floral(q)
Butyric acid	Butanoic acid	**60**,73,88	827(g)	1627(d)	Sweaty, cheesy unpleasent(h)
3-Methyl butyric acid	3-Methylbutanoic acid		834(h)	1675(r)	Cheesy, sweaty(h)
Valeric acid	Pentanoic acid	**60**,73,101			Strawberry(q)
Caproic acid	Hexanoic acid	**60**,73	982(g)	1865(d)	Faintly cheesy sweaty(h)
Caprylic acid	Octanoic acid	**60**,73,144	1163(h)		Sweaty, penetrating(h)
Capric acid	Decanoic acid	**60**,73,172			Rancid(p)
Lauric acid	Dodecanoic acid	60,73,200			Soap(f)
Myristic acid	Tetradecanoic acid	43,60,**73**			Coconut oil
Palmitic acid	Hexadecanoic acid	**43**,60,73		2820(d)	

Common name	IUPAC name	m/z fragment	LRI Apolar column	LRI Polar column	Aroma notes
DERIVED FROM CAROTENOIDS					
Norisoprenoids					
β-Ionone	4-(2,6,6-Trimethyl-cyclohex-1-enyl)-but-3-en-2-one	43, **177**,192		1614(c)	Violet (m)
3-Oxo-*a*-ionol	9-Hydroxymegastigma-4,7-dien-3-one	108	1648(a)	2518(b) 2658(d) 1937(c)	
Blumenol C	4-(3-Hydroxybutyl)-3,5,5-trimethyl-cyclohex-2-en-1-one	41,**43**,108		2002(c)	
Blumenol A (vomifoliol)	6,9-Dihydroxymegastigma-4,7-dien-3-one	189,**207**, 224		2180(c)	
Vitispirane	2,6,6-Trimethyl-10-methyliden-1-oxospiro[4,5]dec-7-ene		1286(e)	1327(c)	Floral, fruity, earthy, woody
3-Oxo-retro-*a*-ionol	9-hidroximegastigma-4,6-dien-3-one (*cis* and *trans*)			2001(c) 2081(c) 2797(d)	

(a)DB5 column, Fernandez de Simon et al., 2009 ; (b)Carbowax column, Fernandez de Simon et al., 2009; (c)DB1701 (medium polar) column, Sefton et al., 1990; (d)Stabilwax column, Natali et al., 2006 (e)HP5 column, Jordão et al.,2006; (f)DBwax column, Caldeira et al.,2008; (g)HP5 column, Vichi et al., 2007; (h) SPB1 column, Diaz-Maroto et al., 2008; (i)Rodriguez-Bencomo et al., 2009; (j)Mosedale & Puech, 1998; (k)Piggott et al., 1995; (l)Garde-Cerdán & Ancín-Azpilicueta , 2006a (m)Sáenz-Navajas et al., 2010; (n) Togores, 2004; (o)Gomes da Silva & Chaves das Neves, 1999; (p) Zhao et al, 2011 (q) Cormier et al. 1991; (r)Barata et al., 2011

Table 1. Main volatile compounds formed from wood biopolymer during toasting process. LRI denotes linear retention indexes; *m/z* denotes mass fragment ions and base peak is presented in bold.

The intensity and length of the applied heat affects the production of compounds during macromolecules degradation and define the toasting levels. Designations like untoasted, light, medium and heavy toast are common, but there is no industry standard for toast level.

According to Vivas et al. (1991) the quality and quantity of each volatile compound are strongly related with toasting intensity, but particular characteristics of each species can also determine the rate of modification during toasting process. For instance, when comparing volatile composition of *Q. pyrenaica* and *Q. petraea* wood chips, the increase amount of compounds with the toasting process was less evident in the Portuguese oak (Jordão et al., 2006). These authors pointed out that physical properties and structure of wood may influence heat conduction and reactions upon heating. Caldeira et al., (2006 b) also found that as toasting intensity increases, concentration of the majority of volatile compounds found in wood matrix rises. In several oak species, as well as in Portuguese chestnut, increasing toasting level led to an increase of furanic aldehydes, volatile phenols, 4-hydroxy-but-2-en-lactone and vanillin (Caldeira et al., 2006 b)

It is also relevant that the response of a wood to a particular seasoning and toasting condition is determined by the size of the wood piece, as it affects their structural properties, and hence, the flavour characteristics. Moreover, each piece size shows different extraction kinetics when in contact with wines (Fernández de Simón et al., 2010 b).

Indeed, when heat is applied to wood, a depolymerization of the lignin takes place producing phenolic aldehydes. Therefore, their concentration increases with toast level. Toasting also produce the cleavage of α–β bonds of cinnamic aldehydes and their thermo-oxidation and thermo-decarboxylation, leading to the formation of dimethoxyphenyl units such as syringol, and in heavy toast, to simple phenols as phenol and *o*-cresol (Nonier et al., 2006).

Toasting is such an important parameter that the characteristic profile of a toasted wood is completely different from the untoasted wood. The particular characteristics of macromolecules, lignin, cellulose, hemicellulose and lipids in each wood have a great influence on the volatiles composition of toasted woods. Fernández de Simón et al., (2009) concluded that toasting led to an increase in almost all the compounds when compared to untoasted woods. These increases were more evident in acacia, chestnut and ash woods, concerning lignin, lipids and carbohydrate derivatives. Cherry and ash woods were found to be richer than toasted oak in lignin derivatives, but much poorer in lipid and carbohydrate derivatives.

3. Analytical methodologies for quantification and identification of volatiles compounds from woods

3.1 Sample preparation methods

Wine aroma is a very complex matrix comprising an enormous variety of compounds of many chemical families. Not all of them are odour active compounds, which imply the necessity of a target strategy in order to isolate the compounds of interest from the aqueous/alcoholic matrix.

Frequently, prior to analysis, samples are submitted to a preliminary preparation step, including isolation and concentration of the target compounds. The more used extraction and enrichment technique are liquid-liquid extraction (LLE), solid-phase extraction (SPE) and headspace, comprising the static/equilibration method (HS) and the dynamic method (purge & trap - P&T). These techniques have been recently reviewed by Costa Freitas et al. The quality of the subsequent analysis and results, depend largely on the isolation procedures. Different preparation methodologies might affect the final extract composition for the very same matrix. Nowadays researchers are focusing their attention not only to the extraction efficiency, of the methods used, but also to sustainable methods and thus more environmental friendly. Considering this, miniaturization has been applied and micro-LLE methods, such as dispersive liquid-liquid extraction (DLLME) (Fariña et al., 2007) or pressurized solvent extraction (Natali et al., 2006, Vichi et al., 2007), solid phase micro extraction methods (SPME) (Cai et al., 2009) and single drop microextraction (SDME)

(Martendal et al., 2007; Sillero et al., 2011) arise as prominent techniques for isolation of compounds in several liquid and solid matrices. Special attention has been given to methods which preclude the use of solvents, such as SPME (Bozalongo et al., 2007) and stir bar sorptive extraction (SBSE) (Marín et al., 2005; Tredoux et al., 2008; Rojas et al., 2009). These methods increase, in the case of gas chromatography (GC), the quality of the resulting chromatogram, allowing the detection of several volatile compounds that would, otherwise, be hidden under the solvent peak. As a definition one might state that preparation methods should be reproducible, and should provide an extract, as much as possible, similar to the original matrix.

Extracts obtained after exhaustive extraction techniques, which include solvent extraction and distillation, normally do not reflect the real matrix aroma profile, since they isolate compounds according, simultaneously, to solubility in the used solvent and volatility, considering the matrix properties (e.g. water or alcohol content). These methods provide a profile that reflects the volatile and semi-volatile composition of matrix rather then "real" aroma compounds of the matrix. Moreover, literature also agree, that only a small fraction of the isolated compounds can be correlated with sensory notes of the matrices (Plutowska & Wardencki, 2008). Consequently HS-based methods are currently the most used ones, since they allow the target isolation of the released volatiles from the matrix.

3.1.2 Extraction methods using solvents

In the analysis of wine aroma, the most used solvent extraction methods are LLE (Ortega et al., 2001; Garde-Cerdán & Ancín-Azpilicueta, 2006b), micro-wave assisted extraction (MAE) (Serradilla & de Castro, 2011), ultrasound assisted extraction (UAE) (Cocito et al., 1995; Fernandes et al., 2003), PLE (Natali et al., 2006; Vichi et al., 2007) and SPE (Morales et al., 2004; Campo et al 2007, Weldegergis et al., 2011). The type and quantity of the used organic solvent varies according to the methodology. In order to perform a classical LLE, eventually, a generous amount of solvent is used, and that is the reason why MAE, UAE and PLE became more popular, since the total amount of solvent can be significantly reduced. Pre-fractionation of the extract can also be subsequently performed (Fernandes et al, 2003) in order to simplify the extract for chromatographic analysis. Nevertheless, in all techniques, the obtained organic extract must be concentrated in order to be chromatographically analyzed. If gas chromatography hyphenated with olfactometry (GC-O) is going to be used, the organic extract should be rinsed with aqueous solutions, at different pH, in order to eliminate the presence of non-volatile compounds which are also extracted during the isolation process (Plutowska & Wardencki, 2008). The final extract has to be free of compounds which can produce artifacts during injection into the hot injector of the GC instrument, such as amino acids or fatty acids. The latter present a prolonged odor in the olfactometric port which can falsify the results of the analytes eluting subsequently. Thus a clean-up with resins should be performed. Before analysis a concentration step is also needed, in order to evaporate the excess of solvent prior to the chromatographic analysis. With this necessary step one can simultaneously promote the evaporation of the more volatile compounds and, eventually, odor components degradation, through oxidation processes, since air contact is almost unavoidable. Thus the final extract will not reproduce the original aroma matrix.

3.1.3 Solvent-less extraction methods

Extraction methods precluding the use of solvents have the main advantage of, in a single step, promoting extraction, isolation and concentration of the target compounds, without any other manipulation. At the same time, since it relies in the headspace of a sample, they will, indeed, mirroring the original sample and thus representing the true olfactory perception. This technique can be used in static or dynamic modes, or at a limit, the sample itself can be submitted to direct analysis without any previous treatment. In static mode, after reaching equilibrium, although not necessarily, sampling takes place and, normally, the more volatile compounds are successfully detected. Less volatile compounds, on the other, and depending on their individual vapour pressure, might be also detected. To overcome this drawback, dynamic sampling can be used. Here, purge & trap techniques, in which the compounds of interest are trapped in a suitable adsorbent are the most used. After sampling the trapped compounds can be directly desorbed, by means of a thermal desorber injector, or an extraction with a suitable solvent, before analysis, can be performed. Except for this latter case, the sample is completely lost after analysis, and no repeats are possible. However, no solvent peak is obtained, allowing detection of the more volatile components that, otherwise, could be co-eluting with the solvent peak. Other drawbacks of the dynamic mode should also be mentioned, namely the impact of the chemical nature of the adsorbent, on the trapped compounds. The most used trap materials are Tenax TA, Porapak Q and Lichrolut EN, since their affinity for ethanol is small (Escudero et al., 2007; Weldegergis et al, 2011). After the introduction of SPME (Belardi & Pawliszyn, 1989; Arthur & Pawliszyn, 1990), most of the HS methods started to use this method to perform aroma analysis. The obvious advantages of SPME are the fact that it does not involve critical sample manipulations. SPME is a simple and clean extraction method, comprising in a single step, all the necessary steps mandatory in aroma analysis: extraction, isolation and concentration (Rocha et al., 2007). In this technique, depending on the fibre physico-chemical properties (chemical nature and thickness), target analysis is, somehow, possible. However if a wide screening of the sample is aimed, more than one type of fibre should be tested. SPME demands a careful optimization of the experimental conditions in order to be used for qualitative and quantitative studies. The efficiency, accuracy and precision of the extraction methodology is directly dependent on extraction time, sample agitation, pH adjustment, salting out, sample and/or headspace volume, temperature of operation, adsorption on container walls and desorption conditions (Pawliszyn, 1997) besides fibre coating and thickness.

3.2 Chromatographic methods

The contact between wines and wood during the fermentation or storage process is a common practice in wine making. In order to improve the organoleptic characteristics of wine, storage in wood based barrels or the addition of wood chips is used in their natural form or after seasoning and thermal treatment by toasting or charring. The wood types for enological purposes comprises cherry, acacia, chestnut, mulberry, and especially, oak. Together with changes in colour and taste, the aroma of the resulting wine product is strongly affected (Garde-Cerdán & Ancín-Azpilicueta, 2006a). The porous nature of wood, promotes oxidation reactions, forming a wide variety of volatile compounds, otherwise not present in the final product. Some compounds are also directly extracted into the wine during the long lasting contact between the wood and the wine. Together with endogenous volatile compounds, they contribute to the final wine bouquet. The most representative compounds, which reflect the wood influence in wine aroma, are furfural derivates, such as furfural, 5-methylfurfural and furfuryl alcohol, phenolic aldehydes and ketones, such as vanillin and acetovanillone and a large variety of volatile phenols, such as guaiacol, 4-ethylphenol, 4-ethylguaiacol, 4-vinylphenol, 4-vinylguaiacol and eugenol and also phenolic acids (Tredoux et al., 2008; Weldergegis et al., 2011). The two oak lactones isomers, *cis*- and *trans*-β-methyl-γ-octalactone and γ-nonalactone are also present, especially if wood has been submitted to a previous seasoning or toasting process (Carrillo et al., 2006). Because they possess a very low sensory threshold, even in very low concentration, they strongly influence the final aroma. Liquid chromatographic (LC) methods, like capillary electrophoresis (CE) (Sádecká & Polonský, 2000), high performance liquid chromatography (HPLC) and ultra performance liquid chromatography (UPLC) are suitable methods to be used in order to separate, detect and quantify the above mentioned chemical classes, as recently reviewed (Cabrita et al., 2008; Kalili & de Villiers, 2011). The strong aromatic ring chromophore allows ease and sensitive detection by ultra-violet-visible (UV) detection devices or by mass spectrometry (MS). UPLC also reached better limits of detection (LOD) in the low ppb range, ten times higher for phenol compounds and furanic derivatives, when compared with HPLC methods (Chinnici et al., 2007). Considering the long-term storage in charred barrels, the presence of polycyclic aromatic hydrocarbons (PAHs) in aged alcoholic beverages has been proposed as an emerging issue. Although the transference rate, to the hydro-alcoholic mixture, of these high hydrophobic compounds is expected to be low, LODs less then 1 ppb are achievable when LC techniques are coupled with fluorimetric detection. Since the majority of the above mentioned compound classes are semi-volatile compounds, together with the endogenous compounds present in the wine, wine becomes an even more complex matrix to be analyzed as a whole. Therefore the analytical methodologies should be sufficiently powerful and sensitive to separate and detect hundreds of compounds of quite different concentrations and volatilities. The advent of gas chromatography (GC) significantly enhanced the systematic study of wine aroma, and its use, particularly when combined with mass spectrometry (MS), contributed significantly to the knowledge of wine volatile composition. Extracts from acacia, chestnut, cherry, mulberry, and oak wood, used in barrels for wine and spirits aging, have already been studied by GC/MS positive ion chemical ionization (PICI). These method, using methane as reagent gas, produce a high yield of the protonated molecular ion of volatile phenols, and allows compounds identification to be confirmed by collision-induced-dissociation (CID) experiments on $[M+H]^+$ species. MS/MS fragmentation patterns can be studied with standard compounds giving more accurate results (Flamini et al., 2007).

Considering that the human nose is still the best detector for aroma active compounds, the hyphenation of GC with olfactometry (GC-O) is mandatory, when sensorial activity, of individual compounds, and the knowledge of the relations between the perceived odour and chemical composition, of the volatile fraction, is searched. In GC-O detection is performed by an educated panel of persons and the qualitative and quantitative evaluation of the odour is carried out for each single compound eluting from the chromatographic column. For a given compound a relationship is established between its sensorial activity at a particular concentration (if above the threshold of sensory detection) and its smell, as well as the determination of the time period of sensory activity and the intensity of the odor. This is possible if in parallel to a conventional detector, e.g. flame ionization detector (FID), MS or even a flame photometric detection (FPD) for sulfur speciation (Chin et al., 2011), an olfactometric port is attached to the instrument in order to determine the compound odor. The splitted flow carries the compounds simultaneously to both detection devices, allowing the comparison between each obtained signals. The mass structural information obtained by MS spectra allows the identification of odour active compounds. The quantitative evaluation of the intensity of the odours can be divided in three groups: detection frequency (where NIF - Nasal Impact Frequency values or SNIF - Surface of Nasal Impact Frequency are measured), dilution to threshold (CHARM method - Combined Hedonic Aroma Response Measurement and AEDA - Aroma Extract Dilution Analysis) and finally direct intensity methods (where the intensity of the stimuli and its duration are measured – OSME - Olfactometry by Finger Span Method). All these methods were carefully revised by Plutowska & Wardencki (2008).

It is easily recognizable that GC is the most appropriate instrumental approach, for wine aroma analysis. However, the enormous quantity of peaks obtained together with the similarity of retention factors of many related components means that component overlap will be the general expectation, meaning that complete separation may be largely unachievable for a given extract. In order to improve analysis, new separation technologies should be embraced. It is desirable to increase peak capacity and thus increasing the probability of locating a greater number of baseline resolved components within the chromatographic run. Together with new stationary phases (more thermal and chemically stable and enantioselective) and faster detection systems, this will largely enhance results quality. Multidimensional chromatographic (MD) methods are emerging as important alternatives to analyze such complex matrix, as wines. Whether using the heart-cutting mode (MD-GC) (Campo et al., 2007; Schmarr et al., 2007; Pons et al., 2008), enantio-MDGC (Darriet et al, 2001; Culleré et al., 2009) or comprehensive two-dimensional GC (2D-GC) approach (GC × GC) (Ryan et al., 2005; Rocha et al., 2007; Torres et al., 2011; Weldegergis et al., 2011), MD methods allow the combination of two or more independent, or nearly independent, separation steps, increasing significantly the separation power of the corresponding one-dimensional (1D-GC) techniques and, therefore the physical separation of compounds in complex samples. When coupled with a preparative fraction collection device (selectable 1D/2D GC–O/MS with PFC) one can perform the enrichment of trace compounds with an high odourific impact (Ochiai & Sasamoto, 2011) in order to eventually proceed to further identification by NMR (Marriott et al., 2009). In order to overcome matrix complexity, a further dimensionality can be coupled to achieve separation efficiency. Methoxypyrazines were analyzed by LC-MDGC/MS (Schmarr et al., 2010a), nevertheless the complexity and price of the instrumental set-up encourages more research in this domain. 2D-LC × LC methods (Cacciola et al., 2007, Dugo et al., 2008; Cesla et al. 2009) are also replacing the traditional 1D-LC separation modes, especially in what phenolic volatile concerns. This technique improves the separation of non-volatile polyphenols (Kivilompolo et al., 2008), and allows ease switch between LC and LC × LC during a single analysis providing a better separation target, only, to the most complex part of the chromatogram.

Additionally to the enhanced separations achieved by 2D methods, comprehensive GC and LC also allow an easy comparison between samples, since the obtained chromatograms can be easily inspected for compound markers in quality control, certification and fraud detection. In fact, the probability of correctly determining a particular sample substance is increased, by the correlation of two or more independent data sets, in complex sample analysis. To generate these two sets of data, analysis has to be performed, individually, in two orthogonal columns phases, e.g. non-polar and a polar column for the case or GC × GC or e.g. exclusion and a reversed-phase column in LC × LC. Although this practice improves the quantitative data for key target components, it is often difficult, and may be tedious, to correlate the retention times for all of the peaks from the individual chromatograms. In any 2D chromatogram retention time data from the two independent stationary phases are immediately available. Thus a 2D chromatogram contains much more information than two independent 1D chromatograms. 2D analysis offers a genuine opportunity to characterize individual components based on retention time data alone, provided that high reproducibility of retention times can be achieved, reducing and eventually precluding, the requirement of MS detection for many samples. The reliability of peak position co-ordinates opens the possibility of fingerprinting comparison or statistical treatment in order to characterize the effects of distinct types of wood, the same wood from different origins, or even aging time (Cardeal et al., 2008; Vaz-Freire, 2009; Schmarr et al., 2010 b; Schmarr & Bernhardt, 2010).

4. Conclusion

Each wood species has a characteristic profile, regarding their volatile composition. Heating during processing of wood to be used in cooperage, specially toasting, has a great influence on biopolymers degradation, thus the resulting compounds. The volatile profile of an untoasted wood can be different from the same wood after toasting. Increasing toasting levels generally leads to an increase of volatile compounds but some can also suffer degradation. Compounds released from wood into wine or spirits depends on several factors such as type of wood, size of alternatives (*e.g.* chips and staves), time of contact, but also depends on the chemical characteristics of the beverage, specially the alcohol content.

Recent development in 2D analytical approaches, together with more sensitive and fast mass spectrometers, allow to gain insight in the volatile fraction of the wine/wood interaction system, aiming wine quality.

5. Acknowledgment

Authors wish to thank Fundação para a Ciência e Tecnologia, Ministério da Ciência, Tecnologia e Ensino Superior and Programa Operacional Ciência e Inovação for financial support PTDC/QUI-QUI/100672/2008.

6. References

Ancín-Azpilicueta, C.; Garde-Cerdán T.; Torrea, D. & Jimenez, N. (2004). Extraction of volatile compounds in model wine from different oak woods: Effect of SO_2. *Food Research International,* Vol. 37, pp. 375-383.

Arapitsas, P.; Antonopoulos, A.; Stefanou, E. & Dourtoglou, V.G.(2004). Artificial aging of wines using oak chips. *Food Chemistry,* Vol.4, pp. 563–570

Arthur, C. & Pawliszyn, J. (1990). Solid Phase Microextraction with Thermal Desorption Using Fused Silica Optical Fibers. *Analytical Chemistry,* Vol.62, pp. 2145-2148

Barata, A.; Campo, E; Malfeito-Ferreira, M.; Loureiro, V; Cacho, J. & Ferreira, V. (2011). Analytical and Sensorial Characterization of the Aroma of Wines Produced with Sour Rotten Grapes Using GC-O and GC-MS: Identification of Key Aroma Compounds. *Journal of Agricultural and Food Chemistry,* Vol. 59, pp. 2543–2553

Bautista-Ortín, A.B.; Lencina, A.G.; Cano-López, M.; Pardo-Minguez, F.; López-Roca, J.M. & Gómez-Plaza, E. (2008). The use of oak chips during the ageing of a red wine in stainless steel tanks or used barrels: Effect of the contact time and size of the oak chips on aroma compounds. *Australian Journal of Grape and Wine Research,* Vol. 14, pp. 63-70.

Belardi, R. P. & Pawliszyn, J. B. (1989). The Application of Chemically Modified Fused Silica Fibers in the extraction of Organics from Water Matrix Samples and their Rapid Transfer to capillary columns. *Water Pollution Research Journal of Canada,* Vol.24, pp. 179-189

Bozalongo, R.; Carrillo, J. D.; Torroba, M. A. F. & Tena, M. T. (2007). Analysis of French and American oak chips with different toasting degrees by headspace solid-phase microextraction-gas chromatography–mass spectrometry. *Journal of Chromatography A,* Vol.1173, pp. 10–17

Cabrita, M.J.; Torres, Mª.; Palma, V.; Alves, E.; Patão, R.; Costa Freitas, A. Mª. (2008). Impact of Malolactic Fermentation on Low Molecular Weight Phenolic Compounds. *Talanta,* Vol. 74, pp. 1281–1286

Cacciola, F.; Jandera, P.; Hajduá, Z.; Cěsla, P. & Mondello, L. (2007). Comprehensive two-dimensional liquid chromatography with parallel gradients for separation of phenolic and flavone antioxidants. *Journal of Chromatography A,* Vol.1149, pp. 73–87

Cadahía, E.; Fernández de Simón, B.; & Jalocha, J. (2003). Volatile compounds in Spanish, French, and American oak woods after natural seasoning and toasting. *Journal of Agricultural and Food Chemistry,* Vol. 51, pp. 5923–5932.

Cadahía, E.; Muñoz, L.; Fernández de Simón, B. & García-Vallejo, M. C. (2001). Changes in low molecular weight phenolic compounds in Spanish, French, and American oak woods during natural seasoning and toasting. *Journal of Agricultural and Food Chemistry,* Vol. 49, pp. 1790–1798.

Cai, L.; Koziel, J. A.; Dharmadhikarib, M. & van Leeuwenc, J. (H). (2009). Rapid determination of *trans*-resveratrol in red wine by solid-phase microextraction with on-fiber derivatization and multidimensional gas chromatography–mass spectrometry. *Journal of Chromatography A,* Vol.1216, pp. 281–287

Caldeira, I.; Anjos, O.; Portal, V.; Belchior, A.P. & Canas, S. (2010). Sensory and chemical modifications of wine-brandy aged with chestnut and oak wood fragments in comparison to wooden barrels. *Analytica Chimica Acta,* Vol. 660, pp. 43–52

Caldeira, I.; Bruno de Sousa, R.; Belchior, A.P. & Clímaco, M.C. (2008). A sensory and chemical approach to the aroma of wooden aged Lourinha wine brandy. *Journal of Viticulture and Enology*, Vol.23, pp.97-110

Caldeira, I.; Pereira, R.; Clímaco, M.C.; Belchior, A.P. & Bruno de Sousa, R. (2004). Improved method for extraction of aroma compounds in aged brandies and aqueous alcoholic wood extracts using ultrasound. *Analytica Chimica Acta*, Vol. 513, pp. 125–134

Caldeira, I.; Clímaco, M.C.; Bruno de Sousa, R. & Belchior A.P. (2006 b) Volatile composition of oak and chestnut woods used in brandy ageing: modification induced by heat treatment. *Journal of Food Engineering*, Vol. 76, pp. 202-211

Caldeira, I.; Mateus, A.M. & Belchior A.P. (2006 a). Flavour and odour profile modifications during the first five years of Lourinhâ brandy maturation on different wooden barrels. *Analytica Chimica Acta*, Vol. 563, pp. 265-273

Campo, E.; Cacho, J. & Ferreira, V. (2007). Solid phase extraction, multidimensional gas chromatography mass spectrometry determination of four novel aroma powerful ethyl esters. Assessment of their occurrence and importance in wine and other alcoholic beverages. *Journal of Chromatography A*, Vol.1140, pp. 180–188

Cardeal, Z. L.; de Souza, P. P.; Gomes da Silva, M. D. R. & Marriott, P. J. (2008). Comprehensive two-dimensional gas chromatography for fingerprint pattern recognition in *cachaça* production. *Talanta* Vol.74, pp. 793–799

Carrillo, J.D.; Garrido-López, A. & Tena, M.T. (2006) Determination of volatile oak compounds in wine by headspace solid-phase microextraction and gas chromatography-mass spectrometry *Journal of Chromatography A*, Vol. 1102, pp. 25-36

Cĕsla, P. Hájek, T. & Jandera, P. (2009). Optimization of two-dimensional gradient liquid chromatography separations. *Journal of Chromatography A*, Vol. 1216, pp. 3443–3457

Chatonnet, P. & Dubourdieu, D. (1998) Comparative study of the characteristics of American white oak (*Quercus alba*) and European oak (*Quercus petraea* and *Q. robur*) for production of barrels used in barrel aging of wines. *American Journal of Enology and Viticulture*, Vol 49, pp. 79-85.

Chatonnet, P. ; Cutzach, I. ; Pons, M. & Dubourdieu, D. (1999). Monitoring toasting intensity of barrels by chromatographic analysis of volatile compounds from toasted oak wood. *Journal of Agricultural and Food Chemistry*, Vol. 47, pp. 4310–4318.

Chin, S. T.; Eyres, G. T. & Marriott, P. J. (2011). Identification of potent odourants in wine and brewed coffee using gas chromatography-olfactometry and comprehensive two-dimensional gas chromatography. *Journal of Chromatography A. Vol. 1218, pp.7487-7498*

Chinnici, F.; Natali, N.; Spinabelli, U. & Riponi, C. (2007). Presence of polycyclic aromatic hydrocarbons in woody chips used as adjuvant in wines, vinegars and distillates *LWT*, Vol. 40, pp. 1587–1592

Cormier, F.; Raymond, Y. ; Champagne, C. P. & Morin, A. (1991). Analysis of odor-active volatiles from Pseudomonas fragi grown in milk. *Journal of Agricultural and Food Chemistry*, Vol. 39, pp. 159–161.

Costa Freitas, A. M.; Gomes da Silva, M. D. R. & Cabrita M. J. (accepted). Extraction Techniques and Applications: Food and Beverage. Sampling techniques for the determination of volatile components in grape juice, wine and alcoholic beverages", in *Comprehensive Sampling and Sample Preparation*, J. Pawliszyn (Editor), Elsevier

Cocito, C., Gaetano, G. & Delli, C. (1995). Rapid extraction of aroma compounds in must and wine by means of ultrasound. *Food Chemistry,* Vol.52, pp. 311-320

Culleré, L.; Escudero, A.; Campo, E.; Cacho, J. & Ferreira V. (2009). Multidimensional gas chromatography–mass spectrometry determination of 3-alkyl-2-methoxypyrazines in wine and must. A comparison of solid-phase extraction and headspace solid-phase extraction methods. *Journal of Chromatography A*, Vol.1216, pp. 4040–4045

Cutzach, I. ; Chatonnet, P. ; Henry, R. & Dubourdieu, D. (1997). Identification of volatile compounds with a "toasty" aroma in heated oak used in barrelmaking. *Journal of Agricultural and Food Chemistry*, Vol. 45, pp. 2217–2224.

Darriet, P.; Lamy, S.; La Guerche, S.; Pons, M.; Dubourdieu, D.; Blancard, D.; Steliopoulos, P. & Mosandl, A. (2001). Stereodifferentiation of geosmin in wine *European Food Research & Technology*, Vol.213, pp. 122–125

de Rosso, M.; Cancian, D.; Panighel, A.; Vedova, A.D. & Flamini, R. (2009). Chemical composition released from five different woods used to make barrels for aging wines and spirits: volatile compounds and polyphenols. *Wood Science Technology,* Vol 43, pp. 375-385

Díaz-Maroto, M.C.; Guchu, E.; Castro-Vásquez, L.; de Torres, C. & Pérez-Coello, M.S. (2008) Aroma-active compounds of American, French, Hungarian and Russian oak woods, studied by GC-MS and GC-O. *Flavour and Fragance Journal*, Vol.23, pp. 93-98

Doussot, F. ; De Jeso, B. ; Quideau, S. & Pardon, P. (2002). Extractives content in cooperage oak wood during natural seasoning and toasting; influence of tree species, geographic location, and single-tree effect. *Journal of Agricultural and Food Chemistry*, Vol. 50, pp. 5955–5961.

Dugo, P.; Cacciola, F. Herrero, M.; Donato, P. & Mondello, L. (2008). Use of partially porous column as second dimension in comprehensive two-dimensional system for analysis of polyphenolic antioxidants. *Journal of Separation Science*, Vol.31, pp. 3297–3308

Escalona, H.; Birkmyre, L.; Piggot, J.R. & Paterson, A. (2002) Effect of maturation in small oak casks on the Volatility of red wine aroma compounds. *Analytica Chimica Acta*, Vol. 458, pp. 45-54

Escudero, A.; Campo, E.; Fariña, L.; Cacho, J. & Ferreira, V. (2007). Analytical Characterization of the Aroma of Five Premium Red Wines. Insights into the Role of Odor Families and the Concept of Fruitiness of Wines. *Journal of Agricultural and Food Chemistry*, Vol.55, pp. 4501–4510

Fariña, L.; Boido, E.; Carrau, F. & Dellacassa, E. (2007). Determination of volatile phenols in red wines by dispersive liquid–liquid microextraction and gas chromatography–mass spectrometry detection. *Journal of Chromatography A*, Vol.1157, pp. 46-50

Fernandes, L.; Relva, A. M.; Gomes da Silva, M. D. R. & Costa Freitas, A. M. (2003). Different multidimensional chromatographic approaches applied to the study of wine malolactic fermentation. *Journal of Chromatography A*, Vol. 995, pp. 161–169

Fernández de Simón, B.; Muiño, I. & Cadahía, E. (2010a). Characterization of volatile constituints in commercial oak wood chips. *Journal of Agricultural and Food Chemistry*, Vol. 58, pp. 9587-9596

Fernández de Simón, B.; Cadahía, E.; del Álamo, M. & Nevares, I. (2010b) Effect of size, seasoning and toasting in the volatile compounds in toasted oak wood and in a red wine treated with them. *Analytica Chimica Acta*, Vol. 660 pp. 211–220

Fernández de Simón, B.; Esteruelas, E.; Muñoz, A.M.; Cadahía, E. & Sanz, M. (2009). Volatile Compounds in Acacia, Chestnut, Cherry, Ash, and Oak Woods, with a View to Their Use in Cooperage. *Journal of Agricultural and Food Chemistry*, Vol. 57,pp. 3217–3227

Ferreras, D.; Fernández, E. & Falqué, E. (2002) Note: Effects of oak wood on the aromatic composition of *Vitis vinifera* L. var. *treixadura* wines. *Food Science and Technology International*, Vol. 8, pp. 343-349

Flamini, R.; Vedova, A.D.; Cancian, D.; Panighel, A. & de Rosso, M. (2007) GC/MS-positive ion chemical ionization and MS/MS study of volatile benzene compounds in five different woods used in barrel making. *Journal of Mass Spectrometry*, Vol.42, pp.641-646

Frangipane M.Y.; de Santis, D. & Ceccarelli, A. (2007) Influence of oak woods of different geographical origins on quality of wines aged in barriques and using oak chips. *Food Chemistry*, Vol.103, pp. 46-54

Garde-Cerdán, T., & Ancín-Azpilicueta, C. (2006a). Review of quality factors on wine ageing in oak barrels. *Trends in Food Science &Technology*, Vol.17, pp. 438–447

Garde-Cerdán, T., & Ancín-Azpilicueta, C. (2006b). Effect of oak barrel type on the volatile composition of wine: Storage time optimization. *Lebensmittel-Wissenschaft und-Technologie*,Vol. 39, pp. 199-205

Garde-Cerdán, T.; Rodríguez-Mozaz, S. & Ancín-Azpilicueta, C. (2002). Volatile composition of aged wine in used barrels of French oak and of American oak. *Food Research International*, Vol. 35, pp. 603-610

Garde-Cerdán, T.; Torrea-Goñi, D. & Ancín-Azpilicueta, C. (2004). Acumulation of volatile compounds during ageing of two red wines with different composition. *Journal of Food Engineering*, Vol. 65, pp. 349-356

Giménez-Martínez, R.; López-García de la Serrana, H.; Villalón-Mir, M.: Quesada-Granados, J. & López-Martínez, M. C. (1996). Influence of wood heat treatment, temperature and maceration time on vanillin, syringaldehyde, and gallic acid contents in oak word and wine spirit mixtures. *American Journal of Enology and Viticulture*, Vol. 47, pp. 441–446.

Gomes da Silva, M. D. R. & Chaves das Neves, H. J. (1999). Complementary Use of Hyphenated Purge-and-Trap Gas Chromatography Techniques and Sensory Analysis in the Aroma Profiling of Strawberries (Fragaria ananassa). *Journal of Agricultural and Food Chemistry*, Vol. 47, pp. 4568-4573.

Gómez-Plaza, E.; Pérez-Prieto, L.J.; Fernández-Fernández, J.I. & López-Roca, J. M. (2004). The effect of successive uses of oak barrels on the extraction of oak-related volatile compounds from wine. *International Journal of Food Science & Technology,* Vol. 39, pp. 1069-1078.

Guchu, E.; Díaz-Maroto, M.C.; Pérez-Coello, M.S.; González-Viñas, M.A.; Cabezudo Ibáñez, M.D. (2006) Volatile composition and sensory characteristics of Chardonnay wines treated with American and Hungarian oak chips. *Food Chemistry,* Vol. 99, pp. 350–359

Jarauta, I.; Cacho, J. & Ferreira, V. (2005). Concurrent phenomena contributing to the formation of the aroma of wine during aging in oak wood: An analytical study. *Journal of Agricultural and Food Chemistry,* Vol. 53, pp. 4166-4177

Jiménez-Moreno, N. & Ancín-Azpilicueta, C. (2007). Binding of oak volatile compounds by wine lees during simulation of wine ageing. *Lebensmittel-Wissenschaft und-Technologie,*Vol. 40, pp. 619-624.

Jiménez-Moreno, N.; González-Marco, A. & Ancín-Azpilicueta, C. (2007). Influence of wine turbidity on the accumulation of volatile compounds from the oak barrels. *Journal of Agricultural and Food Chemistry,* Vol. 55, pp. 6244-6251.

Jordão, A.M.; Ricardo da Silva, J.; Laureano, O.; Adams, A.; Demyttenaere, J.; Verhé, R. & De Kimpe, N. (2006). Volatile composition analysis by solid-phase microextraction applied to oak wood used in cooperage (*Quercus pyrenaica* and *Quercus Petraea*): effect of botanical species and toasting process. *Journal of Wood Science,* Vol. 52, pp. 514-521

Kalili, K. M. & de Villiers, A. (2011). Recent developments in the HPLC separation of phenolic compounds. *Journal of Separation Science,* Vol.34, pp. 854–876

Kivilompolo, M.; Obůrka, V. & Hyötyläinen, T. (2008). Comprehensive two-dimensional liquid chromatography in the analysis of antioxidant phenolic compounds in wines and juices. *Analytical & Bioanalytical Chemistry,* Vol.391, pp. 373–380

Madrera, R.R.; Gomis, D.B. & Alonso J.J.M. (2003). Influence of Distillation System, Oak Wood Type, and Aging Time on Volatile Compounds of Cider Brandy. *Journal of Agricultural and Food Chemistry,* Vol. 51, pp. 5709-5714

Marco, J.; Artajona, J.; Larrechi, M. S. & Rius, F. X. (1994). Relationship between geographical origin and chemical composition of wood for oak barrels. *American Journal of Enology and Viticulture,* Vol. 45, pp. 192–200.

Marín, J.; Zalacain, A.; De Miguel, C.; Alonso, G. L. & Salinas, M. R. (2005). Stir bar sorptive extraction for the determination of volatile compounds in oak-aged wines. *Journal of Chromatography A,* Vol.1098, pp. 1–6

Marriott, P. J.; Eyres, G. T. & Dufour, J. –P. (2009). Emerging Opportunities for Flavor Analysis through Hyphenated Gas Chromatography, *Journal of Agricultutal and Food Chemistry,* Vol.57, pp. 9962–9971

Martendal, E.; Budziak, D. & Carasek, E. (2007). Application of fractional factorial experimental and Box-Behnken designs for optimization of single-drop microextraction of 2,4,6-trichloroanisole and 2,4,6-tribromoanisole from wine samples. *Journal of Chromatography A,* Vol. 1148, pp. 131–136

Masson, G.; Guichard, E.; Fournier, N. & Puech, J. L. (1995). Stereoisomers of β-methyl-γ-octalactone. 2. Contents in the wood of French (*Quercus robur* and *Quercus petraea*) and American (*Quercus alba*) oaks. *American Journal of Enology and Viticulture,* Vol. 46, pp. 424- 428.

Masson, G.; Guichard, E.; Fournier, N. & Puech, J. L. (1997) The β-methyl-γ-octalactone stereoisomer contents of European and American oak. Applicable to wines and spirits. *Journal des Sciences et Techniques de la Tonnellerie,* Vol. 3, pp. 9-15.

Morales, M.L.; Benitez, B. & Troncoso, A. M. (2004). Accelerated aging of wine vinegars with oak chips: evaluation of wood flavour compounds. *Food Chemistry,* Vol.88, pp. 305–315

Mosedale, J.R. & Puech, J.-L. (1998) Wood maturation of distelled beverages. *Trends in Food Science & Technology,* Vol. 9, pp. 95-101

Natali, N.; Chinnici, F. & Riponi, C. (2006). Characterization of Volatiles in Extracts from Oak Chips Obtained by Accelerated Solvent Extraction (ASE). *Journal of Agricultural Food Chemistry,* Vol.54, pp. 8190-8198

Netto, C.C.; Moreira, R.F.A. & De Maria, C.A.B. (2003). Note: Volatile profile from Caninha aged in oak (*Quercus* sp) and balsam (*Myroxylon,* sp) barrels. *Food science and Technology International,* Vol. 9, pp. 59-64

Nonier, M.F. ;Vivas, N. ; Vivas de Gaulejac, N. ; Absalon, C.; Soulie, Ph & Fouquet, E. (2006) Pyrolysis-gas chromatography/mass spectrometry of Quercus sp. wood. Application to structural elucidation of macromolecules and aromatic profiles of different species, *Journal of Analytical and Applied Pyrolysis,* Vol. 75, pp. 181–193.

Ochiai. N & Sasamoto, K. (2011). Selectable one-dimensional or two-dimensional gas chromatography–olfactometry/mass spectrometry with preparative fraction collection for analysis of ultra-trace amounts of odor compounds. *Journal of Chromatography A,* Vol.1218, pp. 3180–3185

Ortega, C.; Lopez, R.; Cacho, J. & Ferreira, V. (2001). Fast analysis of important wine volatile compounds: Development and validation of a new method based on gas chromatographic-flame ionization detection analysis of dichloromethane microextracts. *Journal of Chromatography,* Vol.923, pp. 205–214.

Pawliszyn, J. (1997). *Solid Phase Microextraction: Theory and Practice.* Wiley-VCH, Inc., New York, USA, 242 pp..

Pérez-Coello, M. S.; Sanz, J. & Cabezudo, D. (1999) Determination of volatile compounds in hydroalcoholic extracts of French and American oak wood. *American Journal of Enology and Viticulture,* Vol. 50 pp. 162- 165

Pérez-Coello, M.S.; Sánchez, M. A.; García, E.; González-Viñas, M.A.; Sanz, J. &. Cabezudo, M.D. (2000). Fermentation of White Wines in the Presence of Wood Chips of American and French Oak. *Journal of Agricultural and Food Chemistry,* Vol. 48, pp. 885-889

Pérez-Prieto, L.J.; Lopez-Roca, J.M.; Martínez-Cutillas, A.; Pardo-Minguez, F. & Gómez-Plaza, E. (2003). Extraction and formation dynamic of oak-related volatile compounds from different volume barrels to wine and their behavior during bottle storage. *Journal of Agricultural and Food Chemistry,* Vol. 51, pp. 5444-5449.

Pérez-Prieto, L.J.; López-Roca, J.M.; Martínez-Cutillas, A.; Pardo-Minguez, F. & Gómez-Plaza, E. (2002). Maturing wine in oak barrels. Effects of origin, volume, and age of the barrel on the wine volatile composition. *Journal of Agricultural and Food Chemistry,* Vol. 50, pp. 3272-3276.

Piggott, J.R. ; Conner, J.M. & Paterson, A. (1995). *Flavour development in whisky maturation.* in Food Flavors. Generation, Analysis and Process Influence, G. Charalambous (Ed.), pp. 1731 - 1751

Plutowska, B & Wardencki, W. (2008). Application of gas chromatography–olfactometry (GC–O) in analysis and quality assessment of alcoholic beverages - A review, *Food Chem*istry, Vol.107, pp. 449–463

Pons, A.; Lavigne, V.; Eric, F.; Darriet, P. & Dubourdieu, D. (2008). Identification of Volatile Compounds Responsible for Prune Aroma in Prematurely Aged Red Wines. *Journal of Agricultural Food Chemistry,* Vol.56, pp. 5285–5290

Prida, A. & Puech, J.L. (2006) Influence of Geographical Origin and Botanical Species on the Content of Extractives in American, French, and East European Oak Woods. *Journal of Agricultural and Food Chem*istry, Vol. 54, pp. 8115-8126

Puech, J.L. (1987). Extraction of phenolic compounds from oak wood in model solutions and evolution of aromatic aldehydes in wine aged in oak barrels. *American Journal of Enology and Viticulture,* Vol. 38, pp. 236-238

Rocha, S. M.; Coelho, E.; Zrostlíková, J. Delgadillo, I & Coimbra, M. A. (2007). Comprehensive two-dimensional gas chromatography with time-of-flight mass spectrometry of monoterpenoids as a powerful tool for grape origin traceability. *Journal of Chromatography A,* Vol.1161, pp. 292–299

Rodríguez-Bencomo, J.J.; Ortega-Heras, M.; Pérez-Magariño, S. & Gónzalez-Huerta (2009). Volatile compounds of red wines macerated with Spanish, American, and French oak chips. *Journal of Agricultural and Food Chemistry,* Vol. 57, pp. 6383-6391

Rodríguez-Bencomo, J.J.; Ortega-Heras, M.; Pérez-Magariño, S.; Gónzalez-Huerta, C. & Gónzalez-Sanjosé, M.L. (2008). Importance of chip selection and elaboration process on the aromatic composition of finished wines. *Journal of Agricultural and Food Chemistry,* Vol. 56, pp. 5102-5111

Rodríguez-Rodríguez, P. & Gómez-Plaza, E. (2011). Differences in the Extraction of Volatile Compounds from Oak Chips in Wine and Model Solutions *American Journal of Enology and Viticulture,* Vol. 62, pp. 127-132

Rojas, F. S.; Ojeda, C. B. & , Pavón, J. M. C. (2009). A Review of Stir Bar Sorptive Extraction. *Chromatographia,* Vol. 69, pp. S79-S94

Ryan, D.; Watkins, P.; Smith, J.; Allen, M. & Marriott, P. (2005). Analysis of methoxypyrazines in wine using headspace solid phase microextraction with isotope dilution and comprehensive two-dimensional gas chromatography. *Journal of Separation Science,* Vol.28, pp. 1075–1082

Sádecká, J. & Polonský. J. (2000). Electrophoretic methods in the analysis of beverages. *Journal of Chromatography A,* Vol. 880, pp.243–279

Sáenz-Navajas M-P; Campo, E.; Fernández-Zurbano, P.; Valentin, D. & Ferreira, V. (2010). An assessment of the effects of wine volatiles on the perception of taste and astringency in wine. *Food Chemistry,* Vol. 121, pp. 1139–1149.

Schmar, H. -G; Ganß, S.; Koschinski, S.; Fischer, U; Riehle, C.; Kinnart, J.; Potouridis, T. & Kutyrev, M. (2010 a). Pitfalls encountered during quantitative determination of 3-alkyl-2-methoxypyrazines in grape must and wine using gas chromatography–mass spectrometry with stable isotope dilution analysis. Comprehensive two-dimensional gas chromatography–mass spectrometry and on-line liquid chromatography-multidimensional gas chromatography–mass spectrometry as potential loopholes. *Journal of Chromatography A,* Vol. 217, pp. 6769–6777

Schmarr, H.-G & Bernhardt, J. (2010). Profiling analysis of volatile compounds from fruits using comprehensivetwo-dimensional gas chromatography and image processing techniques. *Journal of Chromatography A,* Vol.1217, pp. 565–574

Schmarr, H.-G.; Ganß, S.; Sang, W. & Potouridis, T. (2007). Analysis of 2-aminoacetophenone in wine using a stable isotope dilution assay and multidimensional gas chromatography–mass spectrometry. *Journal of Chromatography A,* Vol.1150, pp. 78–84

Schmarr, H.-G; Bernhardt, J.; Fischer, U.; Stephan, A.; Müller, P. & Durner, D. (2010 b). Two-dimensional gas chromatographic profiling as a tool for a rapid screening of the changes in volatile composition occurring due to microoxygenation of red wines. *Analytica Chimica Acta,* Vol.672, pp. 114–123

Sefton, M.; Francis, I.; Williams, P. (1990). Volatile norisoprenoid compounds as constituents of oak woods used in wine and spirit maturation. *Journal of Agricultural and Food Chemistry,* Vol 38, pp. 2045-2049

Sefton, M.A., Spillman, P.J.; Pocock, K.F.; Francis, I.L. & Williams, P.J. (1993). The influence of oak origin, seasoning, and other industry practices on the sensory characteristics and composition of oak extracts and barrel-aged white wines. *Australian Grapegrower & Winemaker,* Vol. 355, pp. 17-25

Serradilla, J. A. P. & de Castro, M. D. L. (2011). Microwave-assisted extraction of phenolic compounds from wine lees and spray-drying of the extract. *Food Chemistry,* Vol.124, pp. 1652–1659

Sillero, I. M.; Herrador, E. A.; Cárdenas, S. & Valcárcel, M. (2011). Determination of 2,4,6-tricholoroanisole in water and wine samples by ionic liquid-based single-drop microextraction and ion mobility spectrometry. *Analytica Chimica Acta,* Vol.702, pp. 199– 204

Singleton, V.L., & Draper, D.E. (1961). Wood chips and wine treatment: The nature of aqueous alcohol extracts. *American Journal of Enology and Viticulture,* Vol.12, pp.152-158

Togores, J. H. (2004). *Tratado de enologia Tomo 2.* Mundiprensa Ed.

Torres, M.P.; Cabrita, M.J.; Gomes da Silva, M. D. R.; Palma, V. & Costa Freitas, A. M^{a}. (2011).The Impact of the Malolactic Fermentation in the Volatile Composition of Trincadeira Wine Variety. *Journal of Food Biochemistry,* Vol. 35, pp. 898-913.

Tredoux, A.; de Villiers, A.; Májek, P.; Lynen, F.; Crouch, A. & Sandra, P. (2008). Stir Bar Sorptive Extraction Combined with GC-MS Analysis and Chemometric Methods for the Classification of South African Wines According to the Volatile Composition, *Journal of Agricultural and. Food Chemistry,* Vol. 56, pp. 4286–4296

van Jaarsveld, F.P.; Hattingh, S. & Minnaar, P. (2009). Rapid induction of ageing character in brandy products – Part III. Influence of toasting. *South African Journal for Enology & Viticulture,* Vol. 30, pp. 24-37

Vaz-Freire, L. T.; Gomes da Silva, M. D. R. & Costa Freitas, A. M. (2009). Comprehensive two-dimensional gas chromatography for fingerprint pattern recognition in olive oils produced by two different techniques in Portuguese olive varieties *Galega Vulgar, Cobrançosa* e *Carrasquenha. Analytica Chimica Acta,* Vol.633, pp. 263–270

Vichi, S.; Santini, C.; Natali, N.; Riponi, C.; Tamames, E. L. & Buxaderas, S. (2007). Volatile and semi-volatile components of oak wood chips analysed by Accelerated Solvent Extraction (ASE) coupled to gas chromatography–mass spectrometry (GC–MS). *Food Chemistry,* Vol.102, pp. 1260–1269

Vivas, N. Glories, Y.; Doneche, B & Gueho, E. (1991). Observation on the oak wood microflora (Quercus sp) during its natural air drying. Annales des Sciences Naturelles- Botanique et Biologie Vegetale, Vol. 11, pp. 149-153

Waterhouse, A.L., & Towey, J.P. (1994). Oak lactone isomer ration distinguishes between wine fermented in American and French oak barrels. *Journal of Agricultural and Food Chemistry,* Vol. 42, pp. 1971-1974

Weldegergis, B. T.; Croucha, A. M.; Górecki, T. & de Villiers, A. (2011). Solid phase extraction in combination with comprehensive two-dimensional gas chromatography coupled to time-of-flight mass spectrometry for the detailed investigation of volatiles in South African red wines. *Analytica Chimica Acta,* Vol.701, pp. 98– 111

Yokotsuka, K; Matsunaga, M. & Singleton, V.L (1994). Comparision of composition of Koshu white wines fermented in oak barrels and plastic tanks. *American Journal of Enology and Viticulture,* Vol. 45, pp. 11-16

Zhao, Y.P.; Wang, L.; Li, J. M.; Pei G. R. & Liu, Q. S. (2011) Comparison of Volatile Compounds in Two Brandies Using HS-SPME Coupled with GC-O, GC-MS and Sensory Evaluation- *South African Journal of Enology and Viticulture,* Vol. 32, pp. 9-20

10

Indoor Air Monitoring of Volatile Organic Compounds and Evaluation of Their Emission from Various Building Materials and Common Products by Gas Chromatography-Mass Spectrometry

Hiroyuki Kataoka[1], Yasuhiro Ohashi[2],
Tomoko Mamiya[1], Kaori Nami[1], Keita Saito[1],
Kurie Ohcho[1] and Tomoko Takigawa[3]
[1]*School of Pharmacy, Shujitsu University, Nishigawara, Okayama,*
[2]*Department of Health Chemistry, Okayama University Graduate School of Medicine, Dentistry and Pharmaceutical Sciences, Tsushima, Okayama,*
[3]*Department of Public Health, Okayama University Graduate School of Medicine, Dentistry and Pharmaceutical Sciences, Shikata, Okayama,*
Japan

1. Introduction

In recent years, increased numbers of people entering modern buildings complain of various symptoms such as dry mucous membranes and skin; irritation of eyes, nose, and throat; chest tightness; headache; and mental fatigue (Kirkeskov et al., 2009). These nonspecific health problems related to indoor environments are caused by volatile organic compounds (VOCs) emitted from various sources such as building materials (Haghighat et al., 2002; Lee et al., 2005; Claeson et al., 2007; Nicolle et al., 2008; Han et al., 2010; Jia et al., 2010), household materials (Kwon et al. 2008), and combusted materials (Liu et al., 2003; Ye, 2008; Fromme et al., 2009; Kabir & Kim, 2011). VOCs are widely used in many household products and are emitted by paints (Afshari et al., 2003; Wieslander & Norbäck, 2010; Chang et al., 2011), adhesives (Wilke et al., 2004), waxes, solvents, detergents, woods (Jensen et al., 2001; Kirkeskov et al., 2009), and items containing them, including carpets (Katsoyiannis et al., 2008), vinyl flooring (Cox et al., 2001 and 2002), air-conditioners (Tham et al., 2004), newspapers (Caselli et al. 2009), printers and photocopiers (Lee et al., 2006). VOCs emitted by these materials can be classified as primary or secondary. Primary emissions are emissions of non-bound or free VOCs within building materials; these are generally low molecular weight compounds utilized in additives, solvents and unreacted raw materials like monomers. Secondary emissions refer to VOCs that were originally chemically or physically bound, and are usually generated following decomposition, oxidation, chain scission, sorption processes, maintenance, or microbial action, followed by their emission (Pedersen et al., 2003; Lee et al., 2005; Wady & Larsson, 2005; Araki et al., 2009 and 2010).

Indoor air quality (Tumbiolo et al., 2005; Salthammer, 2011) has been assessed in various environments, including non-residential buildings (Abbritti & Muzi, 2006; Bruno et al., 2008; Barro et al., 2009; Massolo et al., 2010), residences (Son et al., 2003; Hippelein, 2004; Sax et al., 2004; Ohura et al., 2006; Yamaguchi et al., 2006; Dodson et al., 2009; Liu et al., 2008; Takigawa et al., 2010; Logue et al., 2011), schools (Adgate et al., 2004a; Sohn et al., 2009), hospitals (Takigawa et al., 2004), stores and restaurants (Vainiotalo et al., 2008; Loh et al., 2009). VOCs are regarded as one of the main causes of "sick building syndrome (SBS)" (Harada et al., 2007; Glas et al., 2008; Takeda et al., 2009), and exposure to high concentrations of VOCs can lead to adverse health effects such as acute and chronic respiratory effects, functional alterations of the central nervous system, mucous and dermal irritations, chromosome aberrations, and cancer (Boeglin et al., 2006; Rumchev et al., 2007; Sarigiannis et al., 2011; Zhou et al., 2011). SBS is a serious problem in Japan, and the Ministry of Health, Labour and Welfare (MHLW) of Japan (2002) has advised that total VOC (TVOC) be limited to 400 μg/m^3. This TVOC value, however, was not based on the possible effects of long-term exposure on chronic toxicity. Furthermore, air concentrations of VOCs are generally lower in the home than in the workplace (Larroque et al., 2006; LeBouf et al., 2010), and symptoms related to these low indoor VOC levels and their emission sources are not sufficiently clear. To systematically evaluate the relationship between indoor air pollution and human exposure to VOCs (Gokhale et al., 2008; Delgado-Saborit et al., 2011), it is important to measure VOCs in indoor environments, to assess their possible sources and to determine the source strengths of VOCs to which humans are exposed during working, commuting and rest times. In this chapter, we describe a sensitive and reliable method for the simultaneous determination of VOCs by gas chromatography-mass spectrometry (GC-MS). Using this method, we measured the VOC levels in indoor air of a new building, and we characterized the VOCs emitted from various building materials and common household products.

2. Experimental

2.1 Reagents

A 1 mg/mL standard solution of 39 VOCs (Table 1) in carbon disulfide (CS_2) was purchased from Kanto Kagaku (Tokyo, Japan). All other chemicals were of analytical-reagent grade.

2.2 Gas chromatography-mass spectrometry

GC-MS analysis was performed using a Shimadzu Model QP-2010 gas chromatograph-mass spectrometer in conjunction with a GCMS solution Ver.2 workstation. A fused-silica capillary column of cross-linked DB-1 (J&W, Folsom, CA, USA: 60 m × 0.25 mm i.d., 1.0 μm film thickness) was used. The GC operating conditions included injection and detector temperatures of 260°C; a column temperature of 40°C for 10 min, increasing to 280°C at 8°C/min; an inlet helium carrier gas flow rate of 1.0 mL/min maintained with an electronic pressure controller; and a split ratio of 10:1. The electro impact (EI)-MS conditions included an ion-source temperature of 200°C; ionizing voltage of 70 eV; and selected ion monitoring (SIM) mode detection for each compound in each time fraction. Selected ions and peak numbers of each VOC are shown in Table 1. The 39 VOCs were separated into 8 functional groups (A-H), and the results obtained by an average of duplicate analyses were reported as the total concentrations of target VOCs in each group.

Peak	Retention time (min)	Selected ion (m/z)	VOCs	Group[1)]	Peak	Retention time (min)	Selected ion (m/z)	VOCs	Group[1)]
1	12-13.5	61	Ethyl acetate	E	20	22.5-25	91	*m*-Xylene + *p*-Xylene	A
2		57	*n*-Hexane	B	21		104	Styrene	A
3		83	Chloroform	C	22		91	*o*-Xylene	A
4	13.5-16	62	1,2-Dichloroethane	C	23		43	*n*-Nonane	B
5		43	2,4-Dimethyl-pentane	B	24	25-28.5	93	α-Pinene	H
6		97	1,1,1-Trichloroethane	C	25		105	1,2,3-Trimethyl-benzene	A
7		56	*n*-Butanol	D	26		43	*n*-Decane	B
8		78	Benzene	A	27		146	*p*-Dichloro-benzene	C
9		117	Carbon tetrachloride	C	28		105	1,2,4-Trimethyl-benzene	A
10	16-19	63	1,2-Dichloro-propane	C	29		68	Limonene	H
11		57	2,2,4-Trimethyl-pentane	B	30	28.5-30.5	41	*n*-Nonanal	F
12		43	*n*-Heptane	B	31		43	*n*-Undecane	B
13		43	Methyli-sobutylketone	G	32		119	1,2,4,5-Tetramethyl-benzene	A
14	19-22.5	91	Toluene	A	33	30.5-40	43	*n*-Decanal	F
15		129	Chlorodibromometh ane	C	34		43	*n*-Dodecane	B
16		43	Butyl acetate	E	35		43	*n*-Tridecane	B
17		43	*n*-Octane	B	36		43	*n*-Tetra-decane	B
18		166	Tetrachloro-ethylene	C	37		57	*n*-Penta-decane	B
19	22.5-25	91	Ethylbenzene	A	38		57	*n*-Hexa-decane	B

[1)] A: aromatic hydrocarbon, B: aliphatic hydrocarbon, C: halocarbon, D: alcohol, E: ester, F: aldehyde, G: ketone, H: terpene.

Table 1. VOCs used in this study

2.3 Sampling and analysis of indoor air VOCs

Indoor air quality in 13 rooms in a newly built hospital was assessed by active air sampling and VOC analysis before the hospital was opened (in March) and after one year later (in May). In addition, indoor air VOC monitoring was performed in another newly constructed hospital and in a newly constructed school (in March). This new hospital was built without using adhesives in all floors and walls. The mean room temperature and relative humidity of the rooms were 14°C and 65%, respectively, in March and 25.0°C and 42%, respectively, in May. Active sampling was performed using charcoal sorbent tubes (glass tubes with two sections, 130 mg in front and 65 mg in back; Shibata Kagaku, Tokyo, Japan) and a sampling pump (SP-208 Dual, GL Science Inc., Tokyo, Japan), using the standard method of the MHLW. To enable measuring maximum indoor chemical concentrations, sampling was performed in a room that had been closed for more than 5 h following ventilation. From when ventilation occurred to sampling, all doors of built-in furniture in the room were open. In the center of the room (more than 1 m from the wall and 1.2-1.5 m above the floor), VOCs were collected from air onto charcoal sorbent tubes in duplicate, at a flow-rate of 0.2 L/min for 0.5 h in newly constructed building (before occupation) and at a flow-rate of 6 L/h for 24 h after occupation for one year. As controls, VOCs in the air were also trapped outside, 2-5 m from the building and 1.2-1.5 m above the ground. All samples were sealed in a container with an activated carbon bed, stored in an insulated container, and shipped to our laboratory. The front charcoal sorbent was desorbed with 1 mL of CS_2 by shaking and standing for 1 h. After centrifugation at 3000 rpm for 1 min, the supernatant CS_2 solution was transferred to an autosampler vial, and 1 μL of this solution was injected into the GC-MS system. Outlines of indoor air sampling and the analytical procedure are illustrated in Fig. 1.

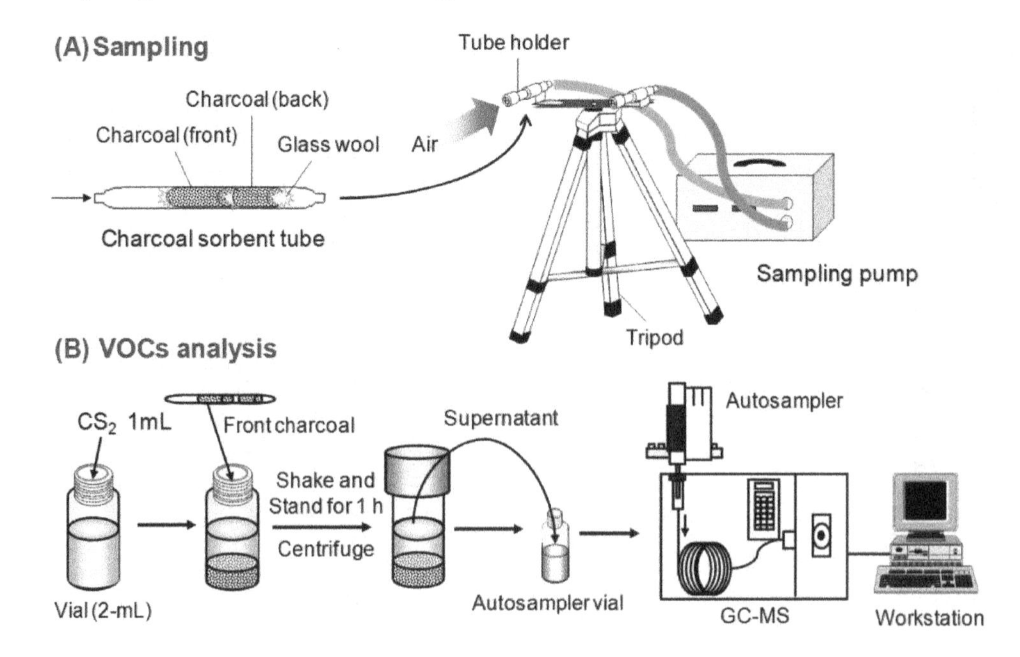

Fig. 1. Outline of indoor air sampling and VOC analysis.

2.4 Sampling and analysis of VOCs emitted from various building materials and common products

VOC emission tests were performed on 16 building materials and 31 common household products, including school supplies, purchased from a local market. Some photographs of these materials are shown in Fig. 2. Woods and hard plastic products were sawn and the sawdust was used for emission tests. Other dry materials, such as carpet, wall paper, newspaper and soft plastic products, were cut finely with scissors or a knife, and the cut pieces were used for emission tests. Wet materials, such as paint, wax, shampoo, glue, paste and ink, were used directly for emission test. Fifteen g of each material were placed in a cleaned small chamber (500-mL volume), and the emitted VOCs were collected onto charcoal sorbent tubes (Shibata Kagaku) by absorption of headspace air using an air sampling pump at a flow-rate of 500 mL/min for 6 h. The adsorbed VOCs on charcoal sorbent were desorbed with 1 mL of CS_2 as described in section 2.3 and analyzed by GC-MS. The VOCs emitted by each material were reported per 180 L. An outline of the emission test is illustrated in Fig. 3.

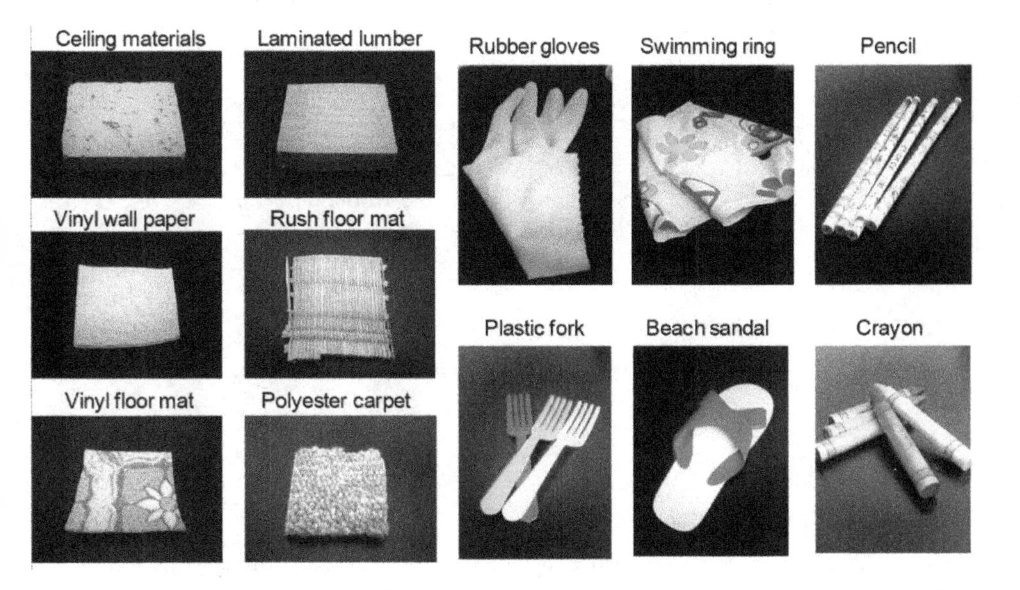

Fig. 2. Several building materials and common products used for emission test.

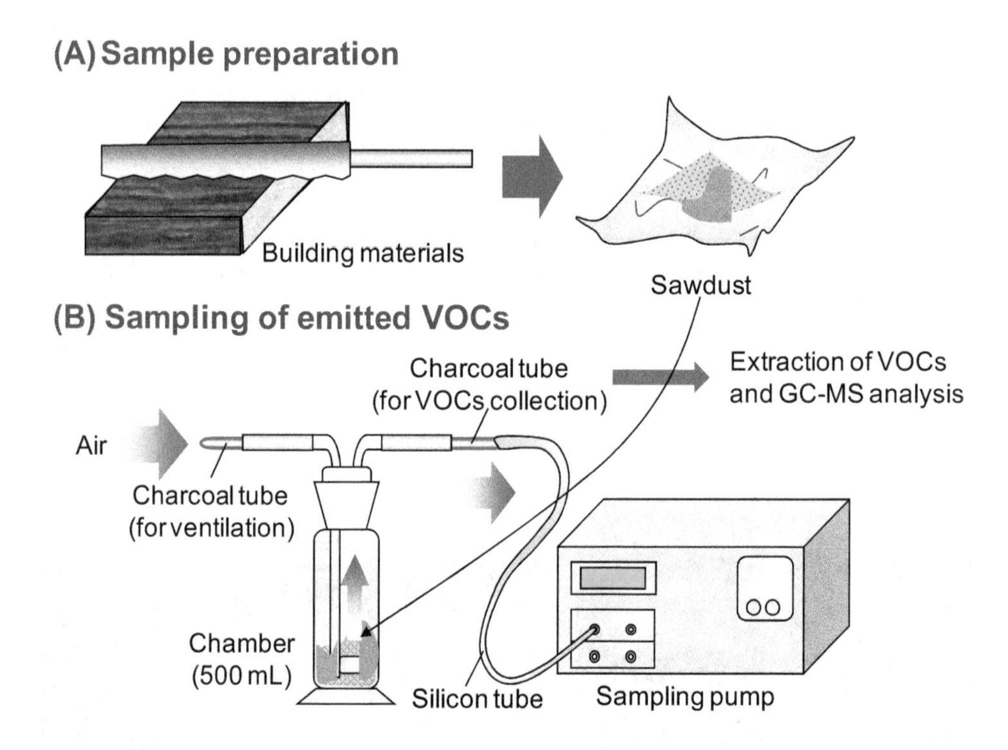

Fig. 3. Outline of emission test.

3. Results and discussion

3.1 GC-MS analysis of VOCs

Mass spectra of VOCs were confirmed by scan mode detection. Although molecular ion peaks of each VOC were observed, the base ion peaks shown in Table 1 were selected for SIM mode detection. The GC-MS method was selective and sensitive, with all 39 VOCs separated on a DB-1 capillary column within 40 min. A typical total ion chromatogram of the VOCs is shown in Fig. 4. The calibration curve for each VOC was linear ($r>0.9992$) in a range from 0.1 to 10 μg/mL CS_2, and the limits of detection (LOD) that gave a signal-to-noise ratio of 3 were 0.4-13.4 ng/mL CS_2 (Table 2).

VOCs	Range [1)] (μg/mL CS_2)	Correlation coefficient	LOD [2)] (ng/mL CS_2)	LOQ [3)] (μg/m^3 for 30 min)
Ethyl acetate	0.1-2.0	0.9993	13.4	21.1
n-Hexane	0.1-2.0	0.9996	4.4	11.5
Chloroform	0.1-2.0	0.9995	4.5	31.3
1,2-Dichloroethane	0.1-2.0	0.9994	2.3	34.7
2,4-Dimethylpentane	0.1-2.0	0.9995	0.7	10.9
1,1,1-Trichloroethane	0.1-2.0	0.9994	8.0	35.7
n-Butanol	0.1-2.0	0.9994	12.4	28.5
Benzene	0.1-2.0	0.9998	2.0	19.4
Carbon tetrachloride	0.1-2.0	0.9992	3.9	44.8
1,2-Dichloropropane	0.1-2.0	0.9992	0.5	25.0
2,2,4-Trimethylpentane	0.1-2.0	0.9996	0.9	11.7
n-Heptane	0.1-2.0	0.9995	0.2	13.7
Methylisobutylketone	0.1-2.0	0.9998	1.3	17.3
Toluene	0.1-10	0.9994	0.4	24.6
Chlorodibromomethane	0.1-2.0	0.9997	1.3	87.5
Butyl acetate	0.1-2.0	0.9997	1.1	19.5
n-Octane	0.1-2.0	0.9994	1.6	15.6
Tetrachloroethylene	0.1-2.0	0.9996	1.8	50.5
Ethylbenzene	0.1-10	0.9996	0.6	27.9
m-Xylene + *p*-Xylene	0.1-10	0.9997	0.7	18.0
Styrene	0.1-2.0	0.9997	0.6	30.4
o-Xylene	0.1-10	0.9997	0.4	29.2
n-Nonane	0.1-2.0	0.9998	1.5	16.1
a-Pinene	0.1-2.0	0.9998	0.8	27.3
1,2,3-Trimethylbenzene	0.1-2.0	0.9998	0.4	31.3
n-Decane	0.1-2.0	0.9996	1.4	13.2
p-Dichlorobenzene	0.1-2.0	0.9998	0.5	42.6
1,2,4-Trimethylbenzene	0.1-2.0	0.9998	0.5	38.6
Limonene	0.1-2.0	0.9998	1.1	25.6
n-Nonanal	0.1-2.0	0.9998	2.8	31.0
n-Undecane	0.1-2.0	0.9995	1.5	17.9
1,2,4,5-Tetramethylbenzene	0.1-2.0	0.9998	0.6	34.7
n-Decanal	0.1-2.0	0.9994	3.9	32.0
n-Dodecane	0.1-2.0	0.9994	1.0	16.7
n-Tridecane	0.1-2.0	0.9992	2.1	17.2
n-Tetradecane	0.1-2.0	0.9994	1.4	18.0
n-Pentadecane	0.1-2.0	0.9996	2.4	19.5
n-Hexadecane	0.1-2.0	0.9995	2.3	20.2

[1)] Range 0.1-2.0 μg/mL (n=12), range 0.1-10 μg/mL (n=18); [2)] S/N=3; [3)] S/N=10.

Table 2. Linearity of calibration, limits of detection and limits of quantitation of target VOCs

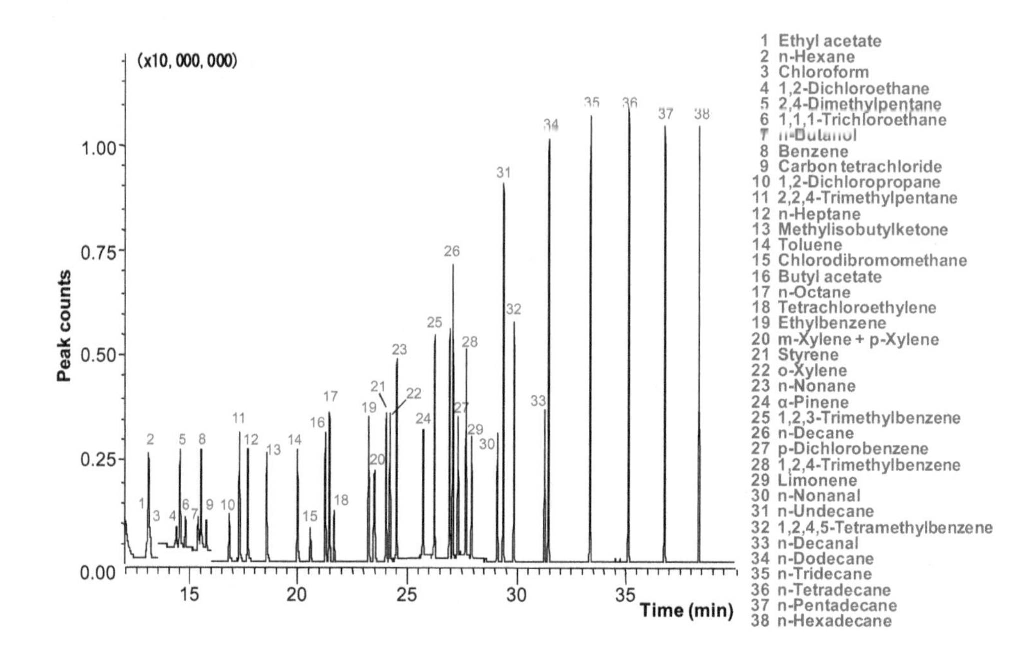

Fig. 4. Typical total ion chromatogram obtained from standard VOCs including 1 µg/mL of each compound.

3.2 Indoor air monitoring of VOCs in newly built buildings

Indoor air VOCs were easily trapped onto charcoal sorbent by the MHLW standard method, with limits of quantitation (LOQ) of VOCs being 10.9-87.5 µg/m³ for 30 min (Table 2). Using this method, we measured the indoor air VOC concentrations in 13 rooms in a newly built, 10 story hospital before occupation and after occupation for one year (Tables 3 and 4). VOC levels varied depending on the presence of indoor building materials, such as paint and furniture. VOCs were not detected, in air sampling obtained once daily from one site outside the hospital. Prior to the building being occupied, aromatic hydrocarbons (toluene, xylenes and ethylbenzene), aliphatic hydrocarbons (mainly *n*-hexane) and esters (ethyl acetate and butyl acetate) were detected with TVOC concentrations exceeding the recommended maximum concentration (400 µg/m³) in 12 of 13 rooms (Fig. 5). Particularly, toluene was detected in all rooms and its concentration exceeded the MHLW recommended maximum concentration (260 µg/m³) in 12 rooms. One year after occupation, however, the TVOC concentrations in the same rooms were below 80 µg/m³, and the indoor levels of toluene and *n*-hexane decreased dramatically, to about 1/100 and 1/60, respectively, of their previous values. **Table 3.** Indoor air VOC amounts in 13 rooms of newly built hospital prior to occupation.

Table 3. Indoor air VOC amounts in 13 rooms of newly built hospital prior to occupation.

VOCs	VOC amounts in indoor air (μg/ m³) [1]/ sampling at 0.2 L/ min for 30 min												
	Guidance room	Doctor office	Radiograph room	10F Lounge	6A Nurse station	6B Nurse station	7A Nurse station	8A Nurse station	8B Nurse station	9A Nurse station	9B Nurse station	10A Nurse station	10B Nurse station
Ethyl acetate	ND[2]	ND	255	ND	ND	ND	ND	ND	ND	ND	ND	ND	ND
n-Hexane	ND	263	ND	ND	ND	111	ND	ND	ND	ND	ND	362	417
Chloroform	ND	ND	ND	ND	ND	ND	ND	ND	ND	ND	ND	ND	ND
1,2-Dichloroethane	ND	ND	ND	ND	ND	ND	ND	ND	ND	ND	ND	ND	ND
2,4-Dimethylpentane	ND	ND	ND	ND	ND	ND	ND	ND	ND	ND	ND	ND	ND
1,1,1-Trichloroethane	ND	ND	ND	ND	ND	ND	ND	ND	ND	ND	ND	ND	ND
n-Butanol	ND	ND	ND	ND	ND	ND	ND	ND	ND	ND	ND	ND	ND
Benzene	ND	ND	ND	ND	ND	ND	ND	ND	ND	ND	ND	ND	ND
Carbon tetrachloride	ND	ND	ND	ND	ND	ND	ND	ND	ND	ND	ND	ND	ND
1,2-Dichloropropane	ND	ND	ND	ND	ND	ND	ND	ND	ND	ND	ND	ND	ND
2,2,4-Trimethylpentane	ND	ND	ND	ND	ND	ND	ND	ND	ND	ND	ND	ND	ND
n-Heptane	ND	ND	ND	ND	ND	ND	ND	ND	ND	ND	ND	ND	ND
Methylisobutylketone	ND	ND	ND	ND	ND	ND	ND	ND	ND	ND	ND	ND	ND
Toluene	218	682	605	640	355	785	1051	420	1733	847	934	850	1494
Chlorodibromomethane	ND	ND	ND	ND	ND	ND	ND	ND	ND	ND	ND	ND	ND
Butyl acetate	ND	ND	ND	ND	618	272	ND	ND	ND	ND	ND	ND	ND
n-Octane	ND	ND	ND	ND	ND	ND	ND	ND	ND	ND	ND	ND	ND
Tetrachloroethylene	ND	ND	ND	ND	ND	ND	ND	ND	ND	ND	ND	ND	ND
Ethylbenzene	ND	186	ND	ND	712	317	160	ND	ND	ND	ND	ND	ND
m-Xylene + *p*-Xylene	ND	ND	ND	ND	744	476	460	ND	ND	ND	ND	ND	ND
Styrene	ND	ND	ND	ND	ND	ND	ND	ND	ND	ND	ND	ND	ND
o-Xylene	ND	ND	ND	ND	250	ND	ND	ND	ND	ND	ND	ND	ND
n-Nonane	ND	ND	ND	ND	ND	ND	ND	ND	ND	ND	ND	ND	ND
α-Pinene	ND	ND	ND	ND	ND	ND	ND	ND	ND	ND	ND	ND	ND
1,2,3-Trimethylbenzene	ND	ND	ND	ND	ND	ND	ND	ND	ND	ND	ND	ND	ND
n-Decane	ND	ND	ND	ND	ND	ND	ND	ND	ND	ND	ND	ND	ND
p-Dichlorobenzene	ND	ND	ND	ND	ND	ND	ND	ND	ND	ND	ND	ND	ND
1,2,4-Trimethylbenzene	ND	ND	ND	ND	ND	ND	ND	ND	ND	ND	ND	ND	ND
Limonene	ND	ND	ND	ND	ND	ND	ND	ND	ND	ND	ND	ND	ND
n-Nonanal	ND	ND	ND	ND	ND	ND	ND	ND	ND	ND	ND	ND	ND
n-Undecane	ND	ND	ND	ND	ND	ND	ND	ND	ND	ND	ND	ND	ND
1,2,4,5-Tetramethylbenzene	ND	ND	ND	ND	ND	ND	ND	ND	ND	ND	ND	ND	ND
n-Decanal	ND	ND	ND	ND	ND	ND	ND	ND	ND	ND	ND	ND	ND
n-Dodecane	ND	ND	ND	ND	ND	ND	ND	ND	ND	ND	ND	ND	ND
n-Tridecane	ND	ND	ND	ND	ND	ND	ND	ND	ND	ND	ND	ND	ND
n-Tetradecane	ND	ND	ND	ND	ND	ND	ND	ND	ND	ND	ND	ND	ND
n-Pentadecane	ND	ND	ND	ND	ND	ND	ND	ND	ND	ND	ND	ND	ND
n-Hexadecane	ND	ND	ND	ND	ND	ND	ND	ND	ND	ND	ND	ND	ND

[1] Mean of duplicate analysis; [2] Not detectable.

VOCs	VOC amounts in indoor air (μg/ m³) [1]/ sampling at 0.2 L/ min for 30 min												
	Guidance room	Doctor office	Radiograph room	10F Lounge	6A Nurse station	6B Nurse station	7A Nurse station	8A Nurse station	8B Nurse station	9A Nurse station	9B Nurse station	10A Nurse station	10B Nurse station
Ethyl acetate	9.0	6.9	7.1	11.7	5.32	11.5	10.8	2.1	ND	3.5	3.1	6.1	8.5
n-Hexane	2.5	ND	1.9	6.0	ND	47.4	22.7	ND	ND	2.3	1.0	1.9	5.9
Chloroform	ND[2]	ND	ND	ND	ND	ND	ND	ND	ND	ND	ND	ND	ND
1,2-Dichloroethane	ND	ND	ND	ND	ND	ND	ND	ND	ND	ND	ND	ND	ND
2,4-Dimethylpentane	ND	ND	0.9	0.8	ND	3.0	1.8	ND	ND	0.5	ND	0.8	1.0
1,1,1-Trichloroethane	ND	ND	ND	ND	ND	ND	ND	ND	ND	ND	ND	ND	ND
n-Butanol	4.2	ND	ND	ND	ND	ND	ND	ND	ND	ND	ND	ND	ND
Benzene	2.7	ND	2.0	2.3	ND	ND	ND	1.2	0.9	1.1	1.1	1.1	2.2
Carbon tetrachloride	ND	ND	ND	ND	ND	ND	ND	ND	ND	ND	ND	ND	ND
1,2-Dichloropropane	ND	ND	ND	ND	ND	ND	ND	ND	ND	ND	ND	ND	ND
2,2,4-Trimethylpentane	ND	ND	ND	ND	ND	ND	ND	ND	ND	ND	ND	ND	ND
n-Heptane	1.1	ND	ND	1.3	ND	ND	ND	ND	ND	ND	ND	ND	1.2
Methylisobutylketone	2.2	2.0	ND	1.6	1.6	1.8	2.1	ND	ND	ND	ND	ND	ND
Toluene	12.9	8.8	8.5	9.8	4.7	8.7	10.0	5.3	5.6	4.6	3.8	7.5	15.6
Chlorodibromomethane	ND	ND	ND	ND	ND	ND	ND	ND	ND	ND	ND	ND	ND
Butyl acetate	2.1	2.8	1.4	1.9	0.7	1.4	1.4	ND	ND	ND	ND	1.5	2.2
n-Octane	ND	ND	ND	1.4	ND	ND	ND	ND	ND	ND	ND	ND	1.3
Tetrachloroethylene	ND	ND	ND	ND	ND	ND	ND	ND	ND	ND	ND	ND	ND
Ethylbenzene	3.4	2.3	3.0	3.3	ND	2.0	1.8	1.3	1.9	1.1	1.5	1.7	4.3
m-Xylene + *p*-Xylene	3.9	3.8	2.5	3.2	ND	1.4	2.1	2.1	2.2	1.7	1.9	2.4	4.6
Styrene	ND	ND	ND	ND	ND	ND	ND	ND	ND	ND	ND	ND	ND
o-Xylene	ND	ND	ND	ND	ND	ND	ND	ND	ND	ND	ND	ND	1.8
n-Nonane	ND	ND	ND	ND	ND	ND	ND	ND	ND	ND	ND	ND	ND
α-Pinene	ND	ND	ND	ND	ND	ND	ND	ND	ND	ND	ND	ND	ND
1,2,3-Trimethylbenzene	ND	ND	ND	ND	ND	ND	ND	ND	ND	ND	ND	ND	ND
n-Decane	ND	ND	ND	ND	ND	ND	ND	ND	ND	ND	ND	ND	1.1
p-Dichlorobenzene	ND	ND	ND	ND	ND	ND	ND	ND	ND	ND	ND	ND	ND
1,2,4-Trimethylbenzene	ND	ND	ND	ND	ND	ND	ND	ND	ND	ND	ND	ND	ND
Limonene	ND	ND	ND	ND	ND	ND	ND	ND	ND	ND	ND	ND	ND
n-Nonanal	ND	ND	ND	ND	ND	ND	ND	ND	ND	ND	ND	ND	ND
n-Undecane	ND	ND	ND	1.5	ND	ND	ND	ND	ND	ND	ND	ND	ND
1,2,4,5-Tetramethylbenzene	ND	ND	ND	ND	ND	ND	ND	ND	ND	ND	ND	ND	ND
n-Decanal	ND	ND	ND	ND	ND	ND	ND	ND	ND	ND	ND	ND	ND
n-Dodecane	ND	ND	ND	ND	ND	ND	ND	ND	ND	ND	ND	ND	ND
n-Tridecane	ND	ND	ND	ND	ND	ND	ND	ND	ND	ND	ND	ND	ND
n-Tetradecane	ND	ND	ND	ND	ND	ND	ND	ND	ND	ND	ND	ND	ND
n-Pentadecane	ND	ND	ND	ND	ND	ND	ND	ND	ND	ND	ND	ND	ND
n-Hexadecane	ND	ND	ND	ND	ND	ND	ND	ND	ND	ND	ND	ND	ND

[1] Mean of duplicate analysis; [2] Not detectable.

Table 4. Indoor air VOC amounts in 13 rooms of newly built hospital after occupation for one year.

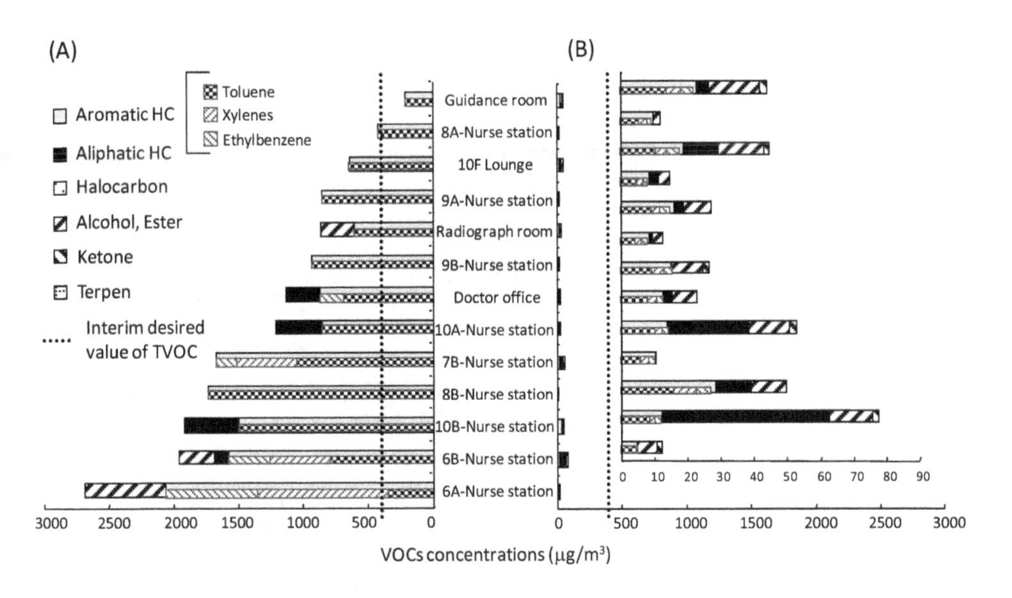

Fig. 5. Comparison of indoor air VOC amounts in 13 rooms of a newly built hospital (A) prior to occupation and (B) after occupation for one year. Air sampling: (A) 0.2 L/min × 30 min and (B) 6 L/h × 24 h.

We also evaluated the relationships among environmental, personal, and occupational factors and changes in the subjective health symptoms in 214 hospital employees (Takigawa et al., 2004). Multiple logistic regression analysis was applied to select variables significantly associated with subjective symptoms that can be induced by SBS. Subjective symptoms of deterioration in the skin, eyes, ears, throat, chest, central nervous system, autonomic system, musculoskeletal system, and digestive system among employees were associated mainly with gender differences and high TVOC concentrations (>1200 μg/m^3). These findings suggest the importance of reducing indoor air VOCs in new buildings to protect employees from the risks of indoor environment-related adverse health effects.

Indoor air VOCs were also measured in unoccupied new buildings, including another newly built hospital that attempted to reduce SBS by not using adhesives in all floors and walls. As shown in Fig. 6A, VOCs were not detected in any rooms or corridors of this hospital. In contrast, TVOC concentrations exceeded the recommended maximum value (400 μg/m^3) in 4 of 10 rooms of a newly built school (Fig. 6B), whereas VOCs were not detected in the other 4 rooms. In 4 rooms, the concentrations of toluene were high, and exceeding the guideline value (260 μg/m^3) of the MHLW. Furthermore, relatively high concentrations of esters (ethyl acetate and butyl acetate) were detected in 4 rooms.

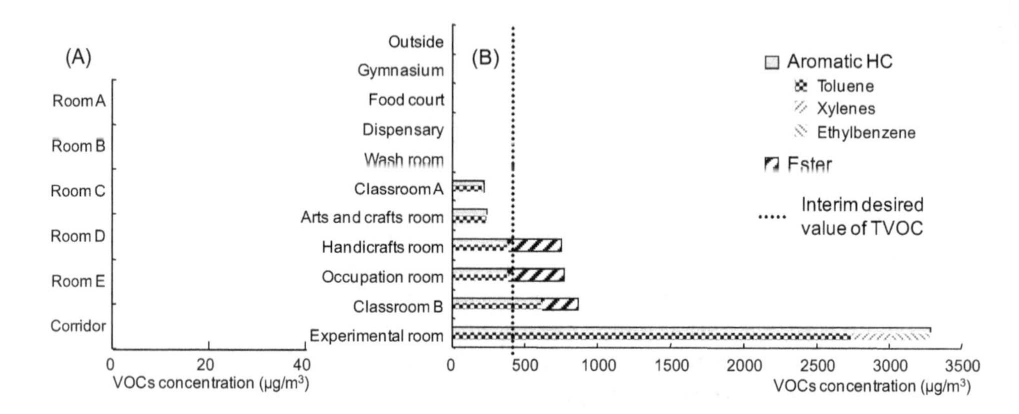

Fig. 6. Indoor air VOC concentrations in rooms of the unoccupied new buildings, (A) a newly built hospital designed to prevent SBS and (B) a newly built school. Air sampling: 0.2 L/min × 30 min

The occurrence and concentrations of VOCs in indoor environments can be affected by outdoor atmospheric conditions, indoor sources, indoor volume, human activities, chemical reactions, ventilation rates, and seasonal factors (Son et al., 2003; Schlink et al., 2004; Massolo et al., 2010). Indoor VOC concentrations have decreased recently in Japan and may be easily reduced by sufficient ventilation and SBS measures. However, measurement of VOC exposure in households with children (Adgate et al., 2004a, b; Sohn et al., 2009) suggested a significant association between VOC exposure and respiratory symptoms such as childhood asthma (Khalequzzaman et al., 2007; Hulin et al., 2010). These findings indicate the necessity of frequent monitoring of VOC exposure in children.

3.3 Emission of VOCs from various building materials and household products

Although various VOCs were detected in newly constructed buildings, they were not detected in the building that took measures to avoid SBS. Therefore, to determine the causal relationship between VOC exposure and SBS onset, it is important to determine the types of building materials and household products that emit VOCs, and the type and quality of VOCs emitted. We therefore collected the VOCs emitted by 16 building materials and 31 household products by a small chamber sampling method (Fig. 3). These emitted VOCs were quantitatively collected onto charcoal sorbent tubes at a flow-rate of 0.5 L/min for 6 h and analyzed by GC-MS.

While there was little emission of VOCs from rush floor mats and ceiling board materials, toluene, chloroform, ethyl acetate, and *n*-hexane were detected in wood chipboard, vinyl wall paper, and vinyl floor mats (Table 5 and Fig. 7). These VOCs may have originated from adhesives and painting materials, which are used to manufacture these products. We found that water-based paints emitted significant amounts of toluene, xylenes, *n*-butanol and high-molecular weight aliphatic hydrocarbons. These quantities emitted may depend on the thickness of the paint layer (Afshari et al., 2003). Some components in these emissions are also highly reactive and may contribute to the health damage.

Table 5. Amounts of VOCs emitted from various building materials.

VOCs	Amounts of VOCs from sample (μg/ 180 L/ 15 g) [1]/ sampling at 0.5 L/ min for 6 h															
	Ceiling board	Gypsum board	Laminated lumber	Plywood	Wood chipboard	Wall paper	Vinyl wall paper (A)	Vinyl wall paper (B)	Rush floor mats	Vinyl floor mats (A)	Vinyl floor mats (B)	Polyester carpets	Floor wax (A)	Floor wax (B)	Adhesive	Water-based paint
Ethyl acetate	ND [2]	1.07	ND	ND	19.59	ND	ND	ND	ND	ND	ND	ND	ND	ND	ND	ND
n-Hexane	ND	1.91	ND	1.17	ND	5.18	0.91	ND	ND	1.51	1.56	1.18	N.D.	5.74	0.56	ND
Chloroform	ND	1.32	ND	1.37	ND	ND	2.80	0.91	ND	ND	ND	3.06	2.35	ND	1.23	ND
1,2-Dichloroethane	ND	ND	ND	ND	ND	ND	ND	ND	ND	ND	ND	ND	ND	ND	ND	ND
2,4-Dimethylpentane	ND	ND	ND	ND	ND	ND	ND	ND	ND	ND	ND	ND	ND	ND	ND	ND
1,1,1-Trichloroethane	ND	ND	ND	ND	ND	ND	ND	ND	ND	ND	ND	ND	ND	ND	ND	ND
n-Butanol	ND	ND	ND	ND	ND	ND	ND	ND	ND	ND	ND	ND	ND	ND	ND	176.55
Benzene	ND	ND	ND	ND	ND	ND	ND	ND	ND	0.28	ND	ND	ND	ND	ND	ND
Carbon tetrachloride	ND	ND	ND	ND	ND	ND	ND	ND	ND	ND	ND	ND	ND	ND	ND	ND
1,2-Dichloropropane	ND	ND	ND	ND	ND	ND	ND	ND	ND	ND	ND	ND	ND	ND	ND	ND
2,2,4-Trimethylpentane	ND	ND	ND	ND	ND	ND	ND	ND	ND	ND	ND	ND	ND	ND	ND	ND
n-Heptane	ND	ND	ND	ND	ND	ND	ND	0.46	ND	ND	ND	ND	ND	ND	ND	8.67
Methylisobutylketone	ND	ND	ND	ND	ND	ND	ND	ND	ND	ND	ND	ND	ND	ND	ND	73.50
Toluene	ND	ND	ND	197	13.91	ND	1.32	0.71	ND	0.93	1.49	0.54	ND	ND	ND	649.64
Chlorodibromomethane	ND	ND	ND	ND	ND	ND	ND	ND	ND	ND	ND	ND	ND	ND	ND	ND
Butyl acetate	ND	ND	ND	ND	ND	ND	ND	ND	ND	ND	ND	ND	ND	ND	ND	ND
n-Octane	ND	ND	ND	ND	ND	ND	ND	ND	ND	ND	ND	ND	ND	ND	ND	ND
Tetrachloroethylene	ND	ND	ND	ND	ND	ND	ND	ND	ND	ND	ND	ND	ND	ND	ND	ND
Ethylbenzene	ND	ND	ND	ND	ND	ND	ND	ND	ND	ND	ND	ND	ND	ND	ND	67.94
m-Xylene + *p*-Xylene	ND	ND	ND	ND	ND	ND	ND	ND	ND	ND	ND	ND	ND	ND	ND	58.89
Styrene	ND	ND	ND	ND	ND	ND	ND	ND	ND	ND	ND	ND	ND	ND	ND	ND
o-Xylene	ND	ND	ND	ND	ND	ND	ND	ND	ND	ND	1.31	ND	ND	ND	ND	23.68
n-Nonane	ND	ND	ND	ND	ND	ND	ND	8.94	ND	ND	ND	ND	ND	ND	ND	27.08
α-Pinene	ND	ND	ND	ND	ND	ND	ND	ND	ND	ND	1.81	9.66	ND	ND	ND	ND
1,2,3-Trimethylbenzene	ND	ND	ND	ND	ND	ND	ND	ND	ND	ND	ND	ND	ND	ND	ND	ND
n-Decane	ND	ND	ND	ND	ND	ND	ND	32.21	ND	ND	ND	ND	ND	ND	ND	88.04
p-Dichlorobenzene	ND	ND	ND	ND	ND	ND	ND	ND	ND	ND	ND	ND	ND	ND	ND	ND
1,2,4-Trimethylbenzene	ND	ND	ND	ND	ND	ND	ND	ND	ND	ND	ND	ND	ND	ND	ND	ND
Limonene	ND	ND	ND	ND	ND	ND	ND	ND	ND	ND	ND	ND	ND	ND	ND	ND
n-Nonanal	ND	ND	ND	ND	ND	ND	ND	ND	ND	ND	ND	ND	ND	ND	ND	ND
n-Undecane	ND	ND	ND	ND	ND	ND	ND	6.38	ND	ND	ND	ND	ND	ND	ND	ND
1,2,4,5-Tetramethylbenzene	ND	ND	ND	ND	ND	ND	ND	4.11	ND	ND	ND	ND	ND	ND	ND	ND
n-Decanal	ND	ND	ND	ND	ND	ND	ND	ND	ND	ND	ND	ND	ND	ND	ND	ND
n-Dodecane	ND	ND	ND	ND	ND	ND	ND	ND	ND	ND	ND	ND	1.41	ND	ND	ND
n-Tridecane	ND	ND	ND	ND	ND	ND	ND	ND	ND	ND	1.68	ND	ND	9.44	ND	ND
n-Tetradecane	ND	ND	ND	ND	ND	ND	ND	1.25	ND	ND	ND	ND	1.56	18.62	ND	ND
n-Pentadecane	ND	ND	ND	ND	ND	ND	ND	ND	ND	ND	ND	ND	ND	27.35	ND	ND
n-Hexadecane	ND	ND	ND	ND	ND	ND	ND	ND	ND	ND	ND	ND	ND	27.59	ND	ND

[1] Mean of duplicate analysis; [2] Not detectable.

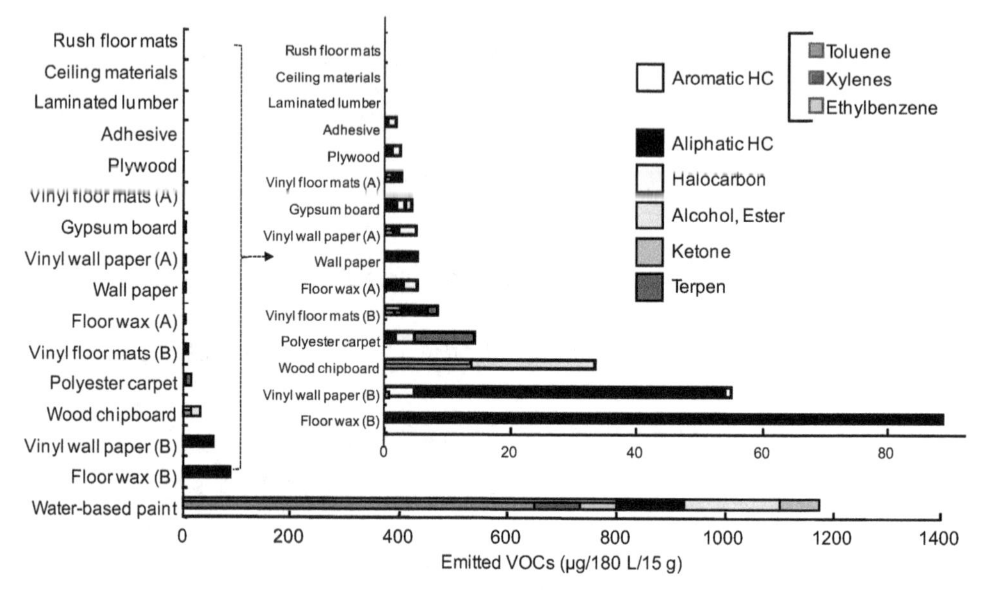

Fig. 7. VOCs emitted from various building materials.

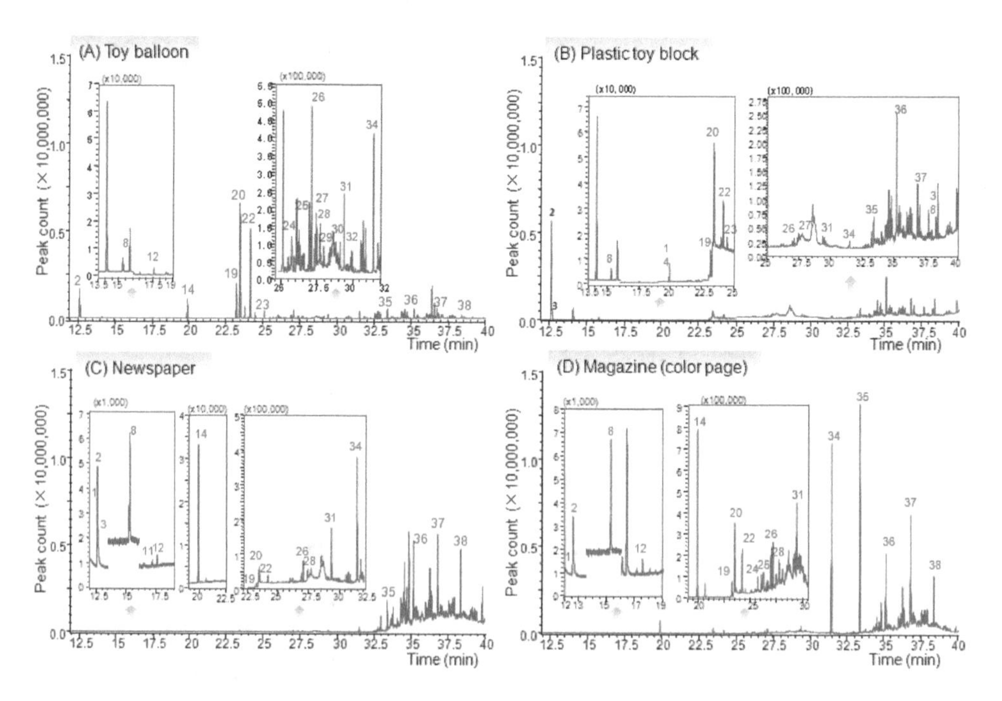

Fig. 8. Typical total ion chromatograms of VOCs emitted from some common products. Peak numbers appear in Fig. 4.

Typical total ion chromatograms of VOCs emitted from some household products are shown in Fig. 8. High concentrations of *n*-hexane, toluene, ethylbenzene and xylenes were detected in a toy rubber balloon (Table 6 and Fig. 9). In addition to toluene, *n*-hexane and chloroform, high concentrations of high-molecular weight aliphatic hydrocarbons were also detected from printed materials such as newspapers and magazines. These are doubtless the main sources of indoor air VOCs at newspaper stands, printing shops, and bookstores (Lee et al., 2006; Barro et al., 2008; Caselli et al., 2009). Evidence has indicated a close relationship between occupational VOC exposure and adverse health effects on workers in the printing industry and in copy centers (Yu et al., 2004). Furthermore, various VOCs were detected in school supplies, including clay, India ink, paint, crayons, glue, and pencils printed with colored paint (Table 7 and Fig. 10). Particularly, paint coating materials are recognized as a major source of VOC exposures (Zhang & Niu, 2002).

These findings may provide semiquantitative estimations of inhalation exposure to VOCs in indoor environments and may allow the selection of safer household products. In particular, the emissions from school supplies are of importance, because they affect the health of children.

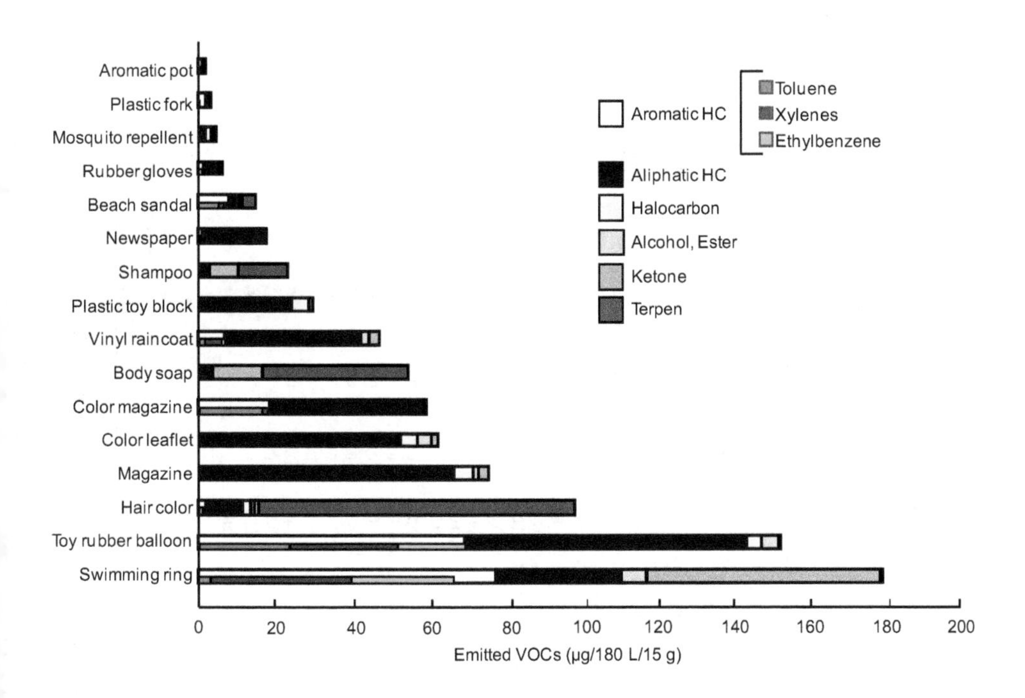

Fig. 9. VOCs emitted from various common products.

VOCs	Amounts of VOCs from sample (μg/ 180 L/ 15 g) [1]/ sampling at 0.5 L/ min for 6 h															
	Newspaper	Magazine	Color magazine	Color leaflet	Plastic toy block	Rubber balloon	Swimming ring	Beach sandal	Vinyl raincoat	Plastic fork	Rubber gloves	Mosquito repellent	Aromatic pot	Shampoo	Body soap	Hair color
Ethyl acetate	0.19	1.87	0.04	2.37	1.03	3.83	1.10	0.07	1.50	0.08	0.28	0.06	0.09	0.07	ND	0.34
n-Hexane	0.06	39.27	0.06	47.71	23.63	70.68	23.59	1.43	33.94	0.72	2.88	0.70	0.28	0.32	0.63	2.73
Chloroform	0.01	4.96	ND	4.39	4.56	3.71	ND	ND	ND	ND	ND	ND	ND	0.12	0.10	0.19
1,2-Dichloroethane	ND [2]	0.19	ND	0.01	ND	ND	ND	ND	ND	ND	ND	ND	ND	ND	ND	ND
2,4-Dimethylpentane	ND	0.39	ND	0.27	ND	ND	ND	ND	ND	ND	ND	ND	ND	ND	ND	ND
1,1,1-Trichloroethane	ND	0.28	ND	0.01	ND	ND	ND	ND	ND	ND	ND	ND	ND	ND	ND	ND
n-Butanol	0.07	1.24	0.03	1.24	ND	0.04	5.57	0.20	0.08	0.00	0.00	0.09	ND	0.04	ND	0.50
Benzene	0.04	0.11	0.06	0.07	0.01	0.01	0.23	0.13	0.31	0.07	0.09	0.08	0.08	0.07	0.07	0.19
Carbon tetrachloride	ND	0.23	ND	0.02	0.04	0.03	0.05	0.01	0.05	ND	ND	ND	ND	0.02	ND	1.84
1,2-Dichloropropane	ND	0.19	ND	ND	ND	0.01	ND	ND	ND	ND	ND	ND	ND	ND	ND	ND
2,2,4-Trimethylpentane	0.01	0.19	0.01	ND	ND	0.01	0.02	ND	ND	ND	ND	ND	ND	ND	ND	ND
n-Heptane	ND	0.18	0.01	0.01	ND	0.06	0.03	ND	0.04	ND	ND	ND	ND	ND	ND	0.08
Methylisobutylketone	ND	0.26	ND	0.01	ND	0.03	57.65	ND	0.06	ND	0.04	ND	0.24	ND	ND	0.09
Toluene	0.64	0.18	16.23	0.05	0.12	23.51	2.86	5.38	1.28	0.05	0.74	0.04	0.10	0.17	0.09	0.93
Chlorodibromomethane	ND	0.12	ND	ND	ND	ND	ND	ND	ND	ND	ND	ND	ND	ND	ND	ND
Butyl acetate	ND	0.29	ND	0.01	ND	0.34	ND	0.39	0.67	0.05	0.52	0.16	0.10	0.02	0.02	0.05
n-Octane	ND	0.14	0.02	0.01	ND	0.27	ND	0.04	0.05	ND	ND	ND	ND	ND	ND	0.23
Tetrachloroethylene	ND	0.07	ND	ND	ND	0.27	ND	ND	ND	ND	ND	ND	ND	ND	ND	ND
Ethylbenzene	0.08	0.11	0.6	0.05	0.08	17.54	26.52	0.34	1.01	0.18	0.20	0.07	0.10	0.07	0.04	0.28
m-Xylene + *p*-Xylene	0.13	0.21	0.99	0.08	0.1	19.93	27.04	1.12	2.71	0.19	0.38	0.11	0.22	0.11	0.08	0.44
Styrene	ND	0.14	0.01	ND	0.01	0.17	0.06	0.03	0.01	1.38	0.01	ND	0.02	0.02	ND	0.01
o-Xylene	0.04	0.11	0.34	0.02	0.03	8.55	10.31	0.33	1.63	0.05	0.11	0.04	0.05	0.03	0.02	0.12
n-Nonane	0.01	0.01	0.03	0.01	0.01	0.56	0.69	0.08	0.05	ND	0.01	ND	0.11	ND	0.01	0.06
α-Pinene	ND	0.09	0.17	ND	ND	0.24	0.13	0.09	0.27	0.01	0.16	0.19	0.28	1.02	0.33	10.54
1,2,3-Trimethylbenzene	0.01	0.11	0.11	0.02	ND	0.1	4.70	0.06	0.05	ND	0.02	0.01	0.16	ND	ND	0.02
n-Decane	0.08	0.05	0.31	0.02	0.01	0.66	2.61	0.19	0.65	0.01	0.03	0.04	ND	0.02	0.04	0.06
p-Dichlorobenzene	ND	0.33	0.01	0.19	0.03	0.24	0.05	0.08	0.12	0.01	0.06	1.47	ND	0.02	0.02	0.07
1,2,4-Trimethylbenzene	0.02	0.14	0.1	0.06	ND	0.12	4.99	0.12	0.04	ND	0.02	0.02	ND	0.05	0.05	0.02
Limonene	0.01	0.15	0.1	0.01	0.01	0.17	0.12	3.76	0.10	0.03	0.48	0.84	ND	1[illegible].80	38.65	72.48
n-Nonanal	0.03	2.84	0.52	1.80	ND	0.06	2.12	0.22	2.39	0.02	0.02	0.03	ND	7.27	12.97	1.16
n-Undecane	0.18	0.49	0.39	0.10	0.01	0.25	3.53	0.21	0.18	0.01	0.02	0.06	ND	0.03	1.40	0.17
1,2,4,5-Tetramethylbenzene	0.01	0.13	0.02	0.03	ND	0.06	1.82	0.04	0.02	ND	0.02	0.01	ND	0.03	0.03	0.01
n-Decanal	ND	0.40	0.03	0.14	ND	ND	2.04	0.22	0.09	ND	ND	0.11	ND	0.29	0.16	0.28
n-Dodecane	0.4	2.55	12.18	0.06	0.01	0.44	1.91	0.25	0.98	0.02	0.03	0.08	ND	0.39	0.53	4.63
n-Tridecane	1.83	1.09	13.62	0.14	0.04	0.61	0.35	0.13	0.12	0.02	0.03	0.17	ND	0.63	0.22	0.28
n-Tetradecane	4.85	5.62	4.56	2.20	0.21	0.57	0.08	0.20	0.49	0.06	0.07	0.26	ND	0.11	0.24	0.97
n-Pentadecane	5.15	14.62	6.77	1.49	0.09	0.25	0.02	0.08	0.05	0.03	0.04	0.08	ND	0.04	0.08	0.23
n-Hexadecane	4.05	4.04	2.96	0.85	0.07	0.11	0.02	0.04	0.08	0.02	0.03	0.05	ND	0.03	0.06	0.10

[1] Mean of duplicate analysis; [2] Not detectable.

Table 6. Amounts of VOCs emitted from various household products.

VOCs	Amounts of VOCs from sample (μg/ 180 L/ 15 g) [1]/ sampling at 0.5 L/ min for 6 h														
	Pencil	Fluid paste	Glue	Adhesive tape	Clay	Rubber band	Indian ink	Stamp ink	Paint (white)	Paint (black)	Paint (red)	Crayon (black)	Crayon (blue)	Crayon (yellow)	Crayon (red)
Ethyl acetate	10.84	0.07	11.13	1.42	1.50	2.31	0.30	ND	0.33	0.27	1.88	0.08	0.07	0.19	3.13
n-Hexane	1.27	0.66	110.16	15.07	33.94	70.52	9.48	0.53	2.25	2.42	36.17	3.12	2.78	6.14	37.33
Chloroform	7.47	ND	0.44	0.03	ND	5.58	4.71	0.10	1.27	1.63	ND	ND	ND	ND	16.17
1,2-Dichloroethane	0.05	ND	0.01	ND	ND	0.02	0.02	0.00	0.01	0.01	ND	ND	ND	ND	0.04
2,4-Dimethylpentane	0.01	ND	ND	0.02	ND	0.02	ND	ND	ND	ND	ND	ND	ND	ND	0.01
1,1,1-Trichloroethane	0.01	ND	0.01	ND	ND	0.01	0.01	ND	ND	ND	ND	ND	ND	ND	0.01
n-Butanol	206.77	ND	0.08	ND	0.08	0.04	ND	0.44	0.07	4.11	0.10	ND	ND	ND	0.44
Benzene	0.08	0.07	0.08	0.01	0.31	0.14	0.08	0.06	0.06	0.09	0.16	0.08	0.08	0.11	0.21
Carbon tetrachloride	0.02	ND	0.02	0.01	0.05	0.04	0.04	ND	0.01	0.01	ND	ND	ND	ND	0.02
1,2-Dichloropropane	ND [2]	ND	ND	ND	ND	ND	ND	ND	ND	ND	ND	ND	ND	ND	ND
2,2,4-Trimethylpentane	ND	ND	0.01	0.03	ND	0.01	ND	ND	ND	ND	ND	ND	ND	ND	ND
n-Heptane	0.01	ND	0.35	8.16	0.04	0.04	ND	ND	ND	0.05	ND	ND	0.04	ND	0.02
Methylisobutylketone	0.08	ND	ND	0.29	0.06	0.06	0.01	ND	ND	0.01	ND	ND	ND	ND	0.01
Toluene	14.81	0.03	0.14	62.35	1.28	1.65	0.12	0.03	0.05	0.10	0.37	0.16	0.15	0.20	0.53
Chlorodibromomethane	ND	ND	ND	ND	ND	ND	ND	ND	ND	ND	ND	ND	ND	ND	ND
Butyl acetate	83.91	0.33	ND	0.68	0.67	0.14	ND	0.54	0.02	0.02	0.03	0.02	0.02	0.04	0.03
n-Octane	84.34	ND	ND	0.64	0.05	0.02	ND	ND	ND	ND	ND	0.03	0.10	0.03	0.05
Tetrachloroethylene	ND	ND	ND	ND	ND	0.01	ND	ND	ND	ND	ND	ND	ND	ND	ND
Ethylbenzene	6.03	0.05	0.05	0.32	1.01	1.11	0.08	0.04	0.04	0.05	0.27	0.13	0.12	0.16	0.31
m-Xylene + *p*-Xylene	7.15	0.07	0.08	0.98	2.71	1.53	0.08	0.07	0.06	0.08	0.39	0.28	0.30	0.32	0.55
Styrene	0.01	ND	ND	ND	0.01	0.02	ND	ND	ND	ND	0.01	ND	ND	ND	ND
o-Xylene	3.10	0.02	0.02	0.01	1.63	0.41	0.02	0.02	0.02	0.03	0.11	0.11	0.13	0.13	0.17
n-Nonane	0.02	ND	ND	ND	0.05	0.03	ND	ND	0.01	0.01	0.01	0.62	0.95	0.69	0.60
α-Pinene	0.03	ND	0.02	ND	0.27	0.09	0.36	ND	ND	0.01	0.03	0.02	0.03	0.03	0.03
1,2,3-Trimethylbenzene	0.19	ND	ND	ND	0.05	0.09	ND	ND	ND	ND	0.01	0.45	0.59	0.49	0.32
n-Decane	0.13	0.01	0.01	ND	0.65	0.10	0.01	0.01	0.01	0.02	0.02	3.85	4.60	4.41	3.08
p-Dichlorobenzene	0.01	ND	0.01	ND	0.12	0.13	ND	0.01	ND	ND	0.02	0.09	0.09	0.16	0.10
1,2,4-Trimethylbenzene	2.23	ND	ND	ND	0.04	0.24	ND	ND	ND	ND	0.01	0.53	0.66	0.58	0.37
Limonene	0.02	0.02	0.04	ND	0.10	0.21	0.02	0.03	ND	0.04	0.02	0.05	0.07	0.06	0.05
n-Nonanal	1.69	ND	5.67	ND	2.39	0.02	0.01	ND	0.01	0.02	0.02	0.55	0.59	0.54	0.24
n-Undecane	0.15	0.01	0.02	ND	0.18	0.16	0.01	0.01	0.01	0.01	0.02	4.32	5.50	5.49	3.58
1,2,4,5-Tetramethylbenzene	2.89	ND	ND	ND	0.02	0.01	ND	ND	ND	ND	ND	0.23	0.29	0.28	0.16
n-Decanal	0.12	ND	0.11	ND	0.09	0.02	1.70	ND	ND	ND	0.05	0.12	ND	9.10	2.80
n-Dodecane	0.08	0.01	0.02	ND	0.98	0.15	0.22	0.01	0.01	0.01	0.02	1.91	2.83	2.80	1.85
n-Tridecane	0.06	0.01	0.05	ND	0.12	0.11	2.08	0.02	0.02	0.04	0.04	0.43	0.70	0.71	0.41
n-Tetradecane	0.08	0.04	0.08	ND	0.49	0.13	0.05	0.04	0.05	0.09	0.09	0.77	1.16	1.36	0.77
n-Pentadecane	0.06	0.02	0.04	ND	0.05	0.07	0.03	0.03	0.04	0.04	0.05	0.18	0.24	0.32	0.16
n-Hexadecane	0.03	0.02	0.03	ND	0.08	0.04	0.02	0.02	0.03	0.03	0.05	0.34	0.43	0.72	0.30

[1] Mean of duplicate analysis; [2] Not detectable.

Table 7. Amounts of VOCs emitted from various school items.

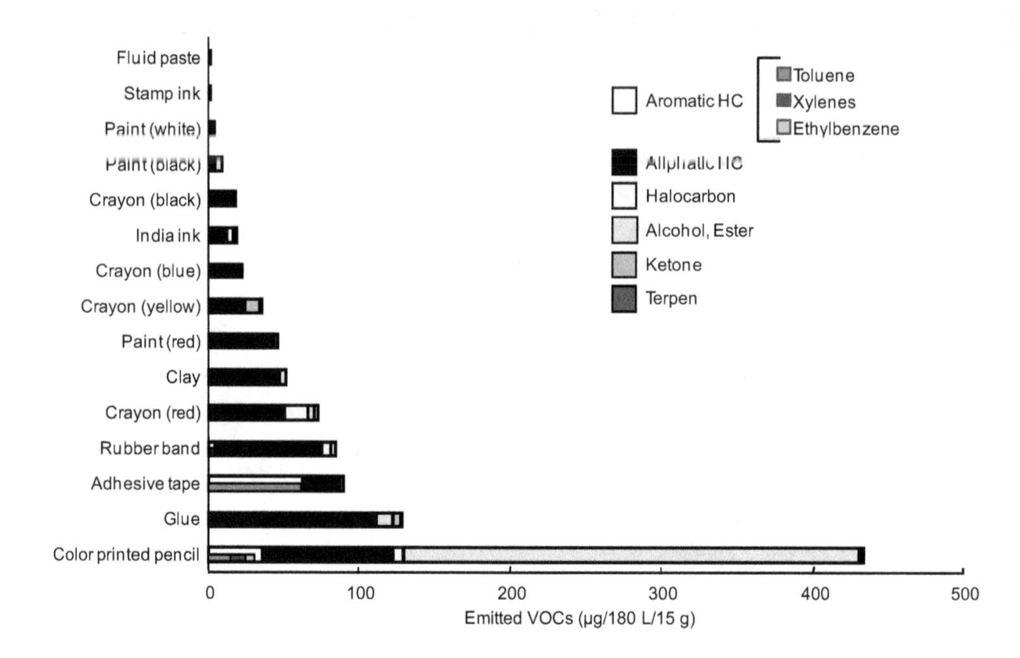

Fig. 10. VOCs emitted from various school items.

4. Conclusion

This chapter provides an analytical method for the determination of VOCs in environmental air samples by GC-MS. This GC-MS method is convenient and reliable, and is useful in evaluating indoor air quality and the sources of VOCs emitted in indoor environments. Indoor air VOC levels in newly constructed buildings exceeded those set by the MHLW. Since humans spend most of their lives indoors, it is necessary to minimize exposure to VOCs affecting human health. Furthermore, we found that various building materials and household products were emission sources of VOCs. Indoor VOC levels associated with these sources can be reduced by increasing outdoor air ventilation, but this entails increased costs in building construction, operation, and energy (Cox et al., 2010). Low VOC-emitting materials are being developed and are used more widely in buildings to help achieve healthier and more productive indoor environments. While VOC-exposure from household products is less than that from building materials, children hypersensitive to these chemicals may be at high risk from directly touching toys and school supplies. Sufficient assessment of the hazards and risks of indoor environments and the regulation of indoor air pollutants such as VOCs are necessary to protect human health, especially children and people who are sensitive to these chemicals. Finally, we hope that this chapter will be beneficial and informative for scientists and students studying environmental pollution and related research fields.

5. Acknowledgment

We are extremely grateful to the late professor Shohei Kira for his valuable instruction. This work was supported by The Ministry of Health, Labour and Walfare of Japan Health Science Research Grant for Research on Environmental Health, The Science Research Promotion Fund, and a Grant-in-Aid for Basic Scientific Research, The Promotion and Mutual Aid Corporation for Private Schools of Japan, and The Yakumo Foundation for Environmental Science.

6. References

Abbritti G. & Muzi G. (2006). Indoor air quality and health in offices and other non-industrial working environments, *Med. Lav.* Vol. 97(2): 410-417.

Adgate J.L., Church T.R., Ryan A.D., Ramachandran G., Fredrickson A.L., Stock T.H., Morandi M.T. & Sexton K. (2004a). Outdoor, indoor, and personal exposure to VOCs in children, *Environ. Health Perspect.* Vol. 112(14): 1386-1392.

Adgate J.L., Eberly L.E., Stroebel C., Pellizzari E.D. & Sexton K. (2004b). Personal, indoor, and outdoor VOC exposures in a probability sample of children, *J. Expo. Anal. Environ. Epidemiol.* Vol. 14: S4-S13.

Afshari A., Lundgren B. & Ekberg L.E. (2003). Comparison of three small chamber test methods for the measurement of VOC emission rates from paint, *Indoor Air* Vol. 13(2):156-165.

Araki A., Eitaki Y., Kawai T., Kanazawa A., Takeda M. & Kishi R. (2009). Diffusive sampling and measurement of microbial volatile organic compounds in indoor air, *Indoor Air* Vol. 19(5): 421-432.

Araki A., Kawai T., Eitaki Y., Kanazawa A., Morimoto K., Nakayama K., Shibata E., Tanaka M., Takigawa T., Yoshimura T., Chikara H., Saijo Y. & Kishi R. (2010). Relationship between selected indoor volatile organic compounds, so-called microbial VOC, and the prevalence of mucous membrane symptoms in single family homes, *Sci. Total Environ.* Vol. 408(10): 2208-2215.

Barro R., Regueiro J., Llompart M. & Garcia-Jares C. (2009). Analysis of industrial contaminants in indoor air: part 1. Volatile organic compounds, carbonyl compounds, polycyclic aromatic hydrocarbons and polychlorinated biphenyls, *J. Chromatogr. A* Vol. 1216(3): 540-566.

Boeglin M.L., Wessels D. & Henshel D. (2006). An investigation of the relationship between air emissions of volatile organic compounds and the incidence of cancer in Indiana countries, *Environ. Res.* Vol. 100: 242-254.

Bruno P., Caselli M., de Gennaro G., Iacobellis S. & Tutino M. (2008). Monitoring of volatile organic compounds in non-residential indoor environments, *Indoor Air* Vol. 18(3): 250-256.

Caselli M., de Gennaro G., Saracino M.R. & Tutino M. (2009). Indoor contaminants from newspapers: VOCs emissions in newspaper stands, *Environ. Res.* Vol. 109(2): 149-157.

Chang Y.M., Hu W.H., Fang W.B., Chen S.S., Chang C.T. & Ching H.W. (2011). A study on dynamic volatile organic compound emission characterization of water-based paints, *J. Air Waste Manag. Assoc.* Vol. 61(1): 35-45.

Claeson A.-S., Sandstrom M. & Sunesson A.-L. (2007). Volatile organic compounds (VOCs) emitted from materials collected from buildings affected by microorganisms, *J. Environ. Monit.* Vol. 9: 240-245

Cox S.S., Little J.C. & Hodgson A.T. (2001). Measuring concentrations of volatile organic compounds in vinyl flooring, *J. Air Waste Manag. Assoc.* Vol. 51(8): 1195-1201.

Cox S.S., Little J.C. & Hodgson A.T. (2002). Predicting the emission rate of volatile organic compounds from vinyl flooring, *Environ. Sci. Technol.* Vol. 36(4): 709-714.

Cox S.S., Liu Z., Little J.C., Howard-Reed C., Nabinger S.J. & Persily A. (2010). Diffusion-controlled reference material for VOC emissions testing: proof of concept, *Indoor Air* Vol. 20(5): 424-433.

Delgado-Saborit J.M., Aquilina N.J., Meddings C., Baker S. & Harrison R.M. (2011). Relationship of personal exposure to volatile organic compounds to home, work and fixed site outdoor concentrations, *Sci. Total Environ.* Vol. 409(3): 478-488.

Dodson R.E., Levy J.I., Houseman E.A., Spengler J.D. & Bennett D.H. (2009). Evaluating methods for predicting indoor residential volatile organic compound concentration distributions, *J. Expo. Sci. Environ. Epidemiol.* Vol. 19(7): 682-693.

Fromme H., Dietrich S., Heitmann D., Dressel H., Diemer J., Schulz T., Jörres R.A., Berlin K. & Völkel W. (2009). Indoor air contamination during a waterpipe (narghile) smoking session, *Food Chem. Toxicol.* Vol. 47(7): 1636-1641.

Glas B., Stenberg B., Stenlund H. & Sunesson A.L. (2008). A novel approach to evaluation of adsorbents for sampling indoor volatile organic compounds associated with symptom reports, *J. Environ. Monit.* Vol. 10(11): 1297-1303.

Gokhale S., Kohajda T. & Schlink U. (2008). Source apportionment of human personal exposure to volatile organic compounds in homes, offices and outdoors by chemical mass balance and genetic algorithm receptor models, *Sci. Total Environ.* Vol. 407(1): 122-138.

Haghighat F., Lee C.S. & Ghaly W.S. (2002). Measurement of diffusion coefficients of VOCs for building materials: review and development of a calculation procedure, *Indoor Air* Vol. 12(2): 81-91.

Han K.H., Zhang J.S., Wargocki P., Knudsen H.N. & Guo B. (2010). Determination of material emission signatures by PTR-MS and their correlations with odor assessments by human subjects, *Indoor Air* Vol. 20(4): 341-354.

Harada K., Hara K., Wei C.N., Ohmori S., Matsushita O. & Ueda A. (2007). Case study of volatile organic compounds in indoor air of a house before and after repair where sick building syndrome occurred, *Int. J. Immunopathol. Pharmacol.* Vol. 20(2 Suppl 2): 69-74.

Hippelein M. (2004). Background concentrations of individual and total volatile organic compounds in residential indoor air of Schleswig-Holstein, Germany, *J. Environ. Monit.* Vol. 6(7): 745-752.

Hulin M., Caillaud D. & Annesi-Maesano I. (2010). Indoor air pollution and childhood asthma: variations between urban and rural areas, *Indoor Air* Vol. 20(6): 502-514.

Jensen L.K., Larsen A., Mølhave L., Hansen M.K. & Knudsen B. (2001). Health evaluation of volatile organic compound (VOC) emissions from wood and wood-based materials, *Arch. Environ. Health.* Vol. 56(5): 419-432.

Jia C., Batterman S., Godwin C., Charles S. & Chin J.Y. (2010). Sources and migration of volatile organic compounds in mixed-use buildings, *Indoor Air* Vol. 20(5): 357-369.

Kabir E. & Kim K.H. (2011). An investigation on hazardous and odorous pollutant emission during cooking activities, *J. Hazard. Mater.* Vol. 188(1-3): 443-454.

Katsoyiannis A., Leva P. & Kotzias D. (2008). VOC and carbonyl emissions from carpets: a comparative study using four types of environmental chambers, *J. Hazard Mater.* Vol. 152(2): 669-676.

Khalequzzaman M., Kamijima M., Sakai K., Chowdhury N.A., Hamajima N. & Nakajima T. (2007). Indoor air pollution and its impact on children under five years old in Bangladesh, *Indoor Air* Vol. 17(4): 297-304.

Kirkeskov L., Witterseh T., Funch L.W., Kristiansen E., Mølhave L., Hansen M.K. & Knudsen B.B. (2009). Health evaluation of volatile organic compound (VOC) emission from exotic wood products, *Indoor Air* Vol. 19(1): 45-57.

Kwon K.D., Jo W.K., Lim H.J. & Jeong W.S. (2008). Volatile pollutants emitted from selected liquid household products, *Environ. Sci. Pollut. Res. Int.* Vol. 15(6): 521-526.

Larroque V., Desauziers V. & Mocho P. (2006). Development of a solid phase microextraction (SPME) method for the sampling of VOC traces in indoor air, *J. Environ. Monit.* Vol. 8: 106-111.

LeBouf R.F., Casteel C. & Rossner A. (2010). Evaluation of an air sampling technique for assessing low-level volatile organic compounds in indoor environments, *J. Air Waste Manag. Assoc.* Vol. 60(2): 156-162.

Lee C.S., Haghighat F. & Ghaly W.S. (2005). A study on VOC source and sink behavior in porous building materials - analytical model development and assessment, *Indoor Air* Vol. 15(3): 183-196.

Lee C.W., Dai Y.T., Chien C.H. & Hsu D.J. (2006). Characteristics and health impacts of volatile organic compounds in photocopy centers, *Environ. Res.* 100(2): 139-149.

Liu J., Guo Y. & Pan X. (2008). Study of the current status and factors that influence indoor air pollution in 138 houses in the urban area in Xi'an, *Ann. N. Y. Acad Sci.* Vol. 1140: 246-255.

Liu W., Zhang J., Hashim J.H., Jalaudin J., Hashim Z. & Goldstein B.D. (2003). Mosquito coil emissions and health implications, *Environ. Health Perspect.* Vol. 111(12): 1454-1460.

Logue J.M., McKone T.E., Sherman M.H. & Singer B.C. (2011). Hazard assessment of chemical air contaminants measured in residences, *Indoor Air* Vol. 21(2): 92-109.

Loh M.M., Houseman E.A., Levy J.I., Spengler J.D. & Bennett D.H. (2009). Contribution to volatile organic compound exposures from time spent in stores and restaurants and bars, *J. Expo. Sci. Environ. Epidemiol.* Vol. 19(7): 660-673.

Massolo L., Rehwagen M., Porta A., Ronco A., Herbarth O. & Mueller A. (2010). Indoor-outdoor distribution and risk assessment of volatile organic compounds in the atmosphere of industrial and urban areas, *Environ. Toxicol.* Vol. 25(4): 339-349.

Ministry of Health, Labour and Walfare of Japan (2002) Committee on sick house syndrome: indoor air pollution progress report no. 4. Ministry of Health, Labour and Walfare of Japan, Tokyo.

Nicolle J., Desauziers V. & Mocho P. (2008). Solid phase microextraction sampling for a rapid and simple on-site evaluation of volatile organic compounds emitted from building materials, *J. Chromatogr. A* Vol. 1208(1-2): 10-15.

Ohura T., Amagai T., Senga Y. & Fusaya M. (2006). Organic air pollutants inside and outside residences in Shimizu, Japan: levels, sources and risks, *Sci. Total Environ.* Vol. 366(2-3): 485-499.

Pedersen E.K., Bjørseth O., Syversen T. & Mathiesen M. (2003). A screening assessment of emissions of volatile organic compounds and particles from heated indoor dust samples, *Indoor Air* Vol. 13(2): 106-117.

Ribes A., Carrera G., Gallego E., Roca X., Berenguer M.A. & Guardino X. (2007). Development and validation of a method for air-quality and nuisance odors monitoring of volatile organic compounds using multi-sorbent adsorption and gas chromatography/mass spectrometry thermal desorption system, *J. Chromatogr. A* Vol. 1140(1-2): 44-55.

Rumchev K., Brown H. & Spickett J. (2007). Volatile organic compounds: do they present a risk to our health?, *Rev. Environ. Health* Vol. 22(1): 39-55.

Salthammer T. (2011). Critical evaluation of approaches in setting indoor air quality guidelines and reference values, Chemosphere Vol. 82(11): 1507-1517.

Sarigiannis D.A., Karakitsios S.P., Gotti A., Liakos I.L. & Katsoyiannis A. (2011). Exposure to major volatile organic compounds and carbonyls in European indoor environments and associated health risk, *Environ. Int.* Vol. 37(4): 743-765.

Sax S.N., Bennett D.H., Chillrud S.N., Kinney P.L. & Spengler J.D. (2004). Differences in source emission rates of volatile organic compounds in inner-city residences of New York City and Los Angeles, *J. Expo. Anal. Environ. Epidemiol.* Vol. 14: S95-S109.

Schlink U., Rehwagen M., Damm M., Richter M., Borte M. & Herbarth O. (2004). Seasonal cycle of indoor-VOCs: comparison of apartments and cites. *Atmos. Environ.* Vol. 38: 1181-1190.

Serrano-Trespalacios P.I., Ryan L. & Spengler J.D. (2004). Ambient, indoor and personal exposure relationship of volatile organic compounds in Mexico City Metropolitan area, *J. Expo. Anal. Environ. Epidemiol.* Vol. 14: S118-S132.

Sohn J., Yang W., Kim J., Son B. & Park J. (2009). Indoor air quality investigation according to age of the school buildings in Korea, *J. Environ. Manage.* Vol. 90(1): 348-354.

Son B., Breysse P. & Yang W. (2003). Volatile organic compounds concentrations in residential indoor and outdoor and its personal exposure in Korea, *Environ. Int.* Vol. 29(1): 79-85.

Takeda M., Saijo Y., Yuasa M., Kanazawa A., Araki A. & Kishi R. (2009). Relationship between sick building syndrome and indoor environmental factors in newly built Japanese dwellings, *Int. Arch. Occup. Environ. Health* Vol. 82(5): 583-593.

Takigawa T., Horike T., Ohashi Y., Kataoka H., Wang, Da-H. & Kira S. (2004). Were volatile organic compounds the inducing factors for subjective symptoms of employees working in newly constructed hospitals?, *Environ. Toxicol.* 19: 280-290.

Takigawa T., Wang B.L., Saijo Y., Morimoto K., Nakayama K., Tanaka M., Shibata E., Yoshimura T., Chikara H., Ogino K. & Kishi R. (2010). Relationship between indoor chemical concentrations and subjective symptoms associated with sick building syndrome in newly built houses in Japan, *Int. Arch. Occup. Environ. Health* Vol. 83(2): 225-235.

Tham K.W., Zuraimi M.S. & Sekhar S.C. (2004). Emission modelling and validation of VOCs' source strengths in air-conditioned office premises, *Environ. Int.* Vol. 30(8): 1075-1088.

Tumbiolo S., Gal J.F., Maria P.C. & Zerbinati O. (2005). SPME sampling of BTEX before GC/MS analysis: examples of outdoor and indoor air quality measurements in public and private sites, *Ann. Chim.* Vol. 95(11-12): 757-766.

Vainiotalo S., Väänänen V. & Vaaranrinta R. (2008). Measurement of 16 volatile organic compounds in restaurant air contaminated with environmental tobacco smoke, *Environ. Res.* Vol. 108(3): 280-288.

Wady L. & Larsson L. (2005). Determination of microbial volatile organic compounds adsorbed on house dust particles and gypsum board using SPME/GC-MS, *Indoor Air* Vol. 15 Suppl 9: 27-32.

Wieslander G. & Norbäck D. (2010). Ocular symptoms, tear film stability, nasal patency, and biomarkers in nasal lavage in indoor painters in relation to emissions from water-based paint, *Int. Arch. Occup. Environ. Health* Vol. 83(7): 733-741.

Wilke O., Jann O. & Brödner D. (2004). VOC- and SVOC-emissions from adhesives, floor coverings and complete floor structures, *Indoor Air* Vol. 14 Suppl 8: 98-107.

Yamaguchi T., Nakajima D., Ezoe Y., Fujimaki H., Shimada Y., Kozawa K., Arashidani K. & Goto S. (2006). Measurement of volatile organic compounds (VOCs) in new residential buildings and VOCs behavior over time, *J. UOEH.* Vol. 28(1): 13-27.

Ye Q. (2008). Development of solid-phase microextraction followed by gas chromatography-mass spectrometry for rapid analysis of volatile organic chemicals in mainstream cigarette smoke, *J. Chromatogr. A* Vol. 1213(2): 239-244.

Yu I.T.S., Lee N.L., Zhang X.H., Chen W.O., Jam Y.T. & Wong T.W. (2004). Occupational exposure to mixtures of organic solvents increases the risk of neurological symptoms among printing workers in Hong Kong, *J. Occup. Environ. Med.* Vol. 46: 323-330.

Zhang L.Z. & Niu J.L. (2002). Mass transfer of volatile organic compounds from painting material in a standard field and laboratory emission cell, *Int. J. Heat & Mass Transfer* Vol. 46: 2415-2423.

Zhou J., You Y., Bai Z., Hu Y., Zhang J. & Zhang N. (2011). Health risk assessment of personal inhalation exposure to volatile organic compounds in Tianjin, China, *Sci. Total Environ.* Vol. 409(3): 452-459.

Permissions

The contributors of this book come from diverse backgrounds, making this book a truly international effort. This book will bring forth new frontiers with its revolutionizing research information and detailed analysis of the nascent developments around the world.

We would like to thank Professor Dr. Mustafa Ali Mohd, for lending his expertise to make the book truly unique. He has played a crucial role in the development of this book. Without his invaluable contribution this book wouldn't have been possible. He has made vital efforts to compile up to date information on the varied aspects of this subject to make this book a valuable addition to the collection of many professionals and students.

This book was conceptualized with the vision of imparting up-to-date information and advanced data in this field. To ensure the same, a matchless editorial board was set up. Every individual on the board went through rigorous rounds of assessment to prove their worth. After which they invested a large part of their time researching and compiling the most relevant data for our readers. Conferences and sessions were held from time to time between the editorial board and the contributing authors to present the data in the most comprehensible form. The editorial team has worked tirelessly to provide valuable and valid information to help people across the globe.

Every chapter published in this book has been scrutinized by our experts. Their significance has been extensively debated. The topics covered herein carry significant findings which will fuel the growth of the discipline. They may even be implemented as practical applications or may be referred to as a beginning point for another development.

The editorial board has been involved in producing this book since its inception. They have spent rigorous hours researching and exploring the diverse topics which have resulted in the successful publishing of this book. They have passed on their knowledge of decades through this book. To expedite this challenging task, the publisher supported the team at every step. A small team of assistant editors was also appointed to further simplify the editing procedure and attain best results for the readers.

Our editorial team has been hand-picked from every corner of the world. Their multi-ethnicity adds dynamic inputs to the discussions which result in innovative outcomes. These outcomes are then further discussed with the researchers and contributors who give their valuable feedback and opinion regarding the same. The feedback is then collaborated with the researches and they are edited in a comprehensive manner to aid the understanding of the subject.

Apart from the editorial board, the designing team has also invested a significant amount of their time in understanding the subject and creating the most relevant covers. They scrutinized every image to scout for the most suitable representation of the subject and create an appropriate cover for the book.

The publishing team has been involved in this book since its early stages. They were actively engaged in every process, be it collecting the data, connecting with the contributors or procuring relevant information. The team has been an ardent support to the editorial, designing and production team. Their endless efforts to recruit the best for this project, has resulted in the accomplishment of this book. They are a veteran in the field of academics and their pool of knowledge is as vast as their experience in printing. Their expertise and guidance has proved useful at every step. Their uncompromising quality standards have made this book an exceptional effort. Their encouragement from time to time has been an inspiration for everyone.

The publisher and the editorial board hope that this book will prove to be a valuable piece of knowledge for researchers, students, practitioners and scholars across the globe.

List of Contributors

Xinghua Guo and Ernst Lankmayr
Institute of Analytical Chemistry and Food Chemistry, Graz University of Technology, Austria

J.H. Sun, D.F. Cui, H.Y. Cai, X. Chen, L.L. Zhang and H. Li
State Key Laboratory of Transducer Technology, Institute of Electronics, Chinese Academy of Sciences, Beijing, China

Vasile Matei, Iulian Comănescu and Anca-Florentina Borcea
Petroleum and Gas University of Ploieşti, Romania

Francis Orata
Masinde Muliro University of Science and Technology, Kenya

Samuel M. Mugo, Lauren Huybregts, Ting Zhou and Karl Ayton
Grant MacEwan University, Edmonton, Canada

S. Pongpiachan
NIDA Center for Research and Development on Disaster Prevention and Management, School of Social and Environmental Development, National Institute of Development, Administration (NIDA), Sereethai Road, Klong-Chan, Bangkapi, Bangkok, Thailand
SKLLQG, Institute of Earth Environment, Chinese Academy of Sciences (IEECAS), Xi'an, China

P. Hirunyatrakul, I. Kittikoon and C. Khumsup
Bara Scientific Co., Ltd., Bangkok, Thailand

Sukesh Narayan Sinha
National Institute of Nutrition (ICMR), Jamia-Osmania, P.O-Hyderabad, India

V. K. Bhatnagar
Principal and Dean, Subharti Institute of Technology & Engineering, Swami Vivekanand Subharti University, Meerut, India

Tetsuo Ohkuwa and Toshiaki Funada
Department of Materials Science and Engineering, Nagoya Institute of Technology, Showa-ku, Nagoya, Japan

Takao Tsuda
Pico-Device Co., Nagoya Ikourenkei Incubator, Chikusa-ku, Nagoya, Japan

Maria João B. Cabrita, Raquel Garcia, Nuno Martins and Ana M. Costa Freitas
School of Science and Technology, Department of Plant Science, Institute of Agricultural and Environmental Science ICAAM Mediterranean, University of Évora, Évora, Portugal

R Marco D.R. Gomes da Silva
EQUIMTE, Department of Chemistry, Faculty of Science and Technology, New University of Lisbon, Campus Caparica, Portugal

Hiroyuki Kataoka, Tomoko Mamiya, Kaori Nami, Keita Saito and Kurie Ohcho
School of Pharmacy, Shujitsu University, Nishigawara, Okayama, Japan

Yasuhiro Ohashi
Department of Health Chemistry, Okayama University Graduate School of Medicine, Dentistry and Pharmaceutical Sciences, Tsushima, Okayama, Japan

Tomoko Takigawa
Department of Public Health, Okayama University Graduate School of Medicine, Dentistry and Pharmaceutical Sciences, Shikata, Okayama, Japan

Printed in the USA
CPSIA information can be obtained
at www.ICGtesting.com
JSHW011410221024
72173JS00003B/496